AF410858

The Ecological Role of Salamanders as Predators and Prey

The Ecological Role of Salamanders as Predators and Prey

Editor

Salvidio Sebastiano

MDPI • Basel • Beijing • Wuhan • Barcelona • Belgrade • Manchester • Tokyo • Cluj • Tianjin

Editor
Salvidio Sebastiano
Università di Genova
Italy

Editorial Office
MDPI
St. Alban-Anlage 66
4052 Basel, Switzerland

This is a reprint of articles from the Special Issue published online in the open access journal *Diversity* (ISSN 1424-2818) (available at: https://www.mdpi.com/journal/diversity/special_issues/salamanders_prey).

For citation purposes, cite each article independently as indicated on the article page online and as indicated below:

LastName, A.A.; LastName, B.B.; LastName, C.C. Article Title. *Journal Name* **Year**, *Volume Number*, Page Range.

ISBN 978-3-0365-3695-8 (Hbk)
ISBN 978-3-0365-3696-5 (PDF)

Contents

About the Editor

Salvidio Sebastiano, Member of the Editorial Board of the MDPI Journal Diversity, is an Associate Professor at the University of Genova (Italy), where he teaches Applied Zoology and Conservation Biology. His general scientific interests are centered on the biology, trophic ecology, and conservation of amphibians and reptiles in the Mediterranean basin. His current research interests focus on the long-term population dynamics of salamanders and the spread of the amphibian chytrid fungi in Italy. He served as President of the Italian Herpetological Society (Societas Herpetologica Italica) from 2004–2008.

Preface to "The Ecological Role of Salamanders as Predators and Prey"

Salamanders are relevant components of many terrestrial and aquatic ecosystems. However, despite the importance of these animals in many resource–consumer networks, their functional role remains remarkably understudied. Therefore, this volume, entitled The Ecological Role of Salamanders as Prey and Predators, provides an opportunity for researchers to highlight the new and innovative research on the ecological role of salamanders and newts in prey–predator systems, their trophic behavior, and the variability of their trophic niche in space and time. Various innovative methods, such as COI metabarcoding and network analysis, are applied in the present study to test both the classical and new hypotheses concerning the trophic ecology of salamanders and their interactions with their prey. The present volume is composed of one review and seven research papers, all of which are published after undergoing a complete and impartial peer-review process. All of the Editors, Reviewers, and Authors that contributed to the realization of this volume are highly commended for their dedicated efforts and commitment to the present publication.

Salvidio Sebastiano
Editor

Editorial

The Ecological Role of Salamanders as Prey and Predators

Sebastiano Salvidio

Department of Earth, Environment and Life Sciences (DISTAV), University of Genova, Corso Europa 26, 16132 Genova, Italy; sebastiano.salvidio@unige.it

Citation: Salvidio, S. The Ecological Role of Salamanders as Prey and Predators. *Diversity* **2022**, *14*, 218. https://doi.org/10.3390/d14030218

Received: 14 March 2022
Accepted: 15 March 2022
Published: 16 March 2022

Publisher's Note: MDPI stays neutral with regard to jurisdictional claims in published maps and institutional affiliations.

Salamanders comprise more than 700 living species, mainly found in the Northern hemisphere (i.e., North and Central America and the northern part of Eurasia), and the Amazon region of South America. Salamanders constitute a diverse clade of amphibians with different reproduction modes that range from completely aquatic to fully terrestrial [1]. Salamanders are key components of many temperate forest ecosystems, in particular in North America [2] and in high altitude lakes, where fish are naturally absent [3]. In these temperate ecosystems, salamanders and newts are top predators that regulate top down the invertebrate prey community [4], while at the same time being high-energetic prey items for birds, mammals and reptiles [5].

In any case, despite their ecological importance, the role of salamanders in resource–consumer networks remains remarkably understudied. Therefore, this Special Issue aims to better understand the different ecological roles of these small vertebrates, both in aquatic and terrestrial ecosystems. Indeed, the eight papers published addressed many of the issues related to the trophic strategies and the trophic position of salamanders in the ecological food web. In particular, one review paper [6] makes a significant contribution to our understanding of salamander and newt populations functioning as predators, competitors and prey in freshwater ecosystems. Furthermore, it appears relevant that four papers were all conducted in underground habitats, both of natural or artificial origin [7–10]. This relatively high number of papers dedicated to the ecological structure and functioning of underground ecosystems clearly indicates a recent growing interest of ecologists and conservation biologists over this highly understudied environment [11,12]. Indeed, in underground habitats, salamanders are, together with cave fish, the only vertebrates that were able to permanently establish reproductive populations. This fact highlights the adaptability of salamanders to extreme subterranean habitats, that were probably colonized to reduce the environmental stress and the predation level experienced in adjacent epigean habitats [13]. Two other papers of this Special Issue analyze the diet of terrestrial salamanders, the first in Spain [14] and the second in Italy [15]. In the former paper, a novel COI metabarcoding approach was used to analyze the dietary habits of the fire salamander *Salamandra salamandra* [14], while in the latter, the authors applied for the first time in salamanders the technique of network analysis to study the trophic strategy of the Alpine salamander *Salamandra atra* [15]. Finally, one paper tested the niche variation hypothesis [16] in a newt community sampled in a complex system of artificial aquatic sites [17]. These authors found that individual specialization was widespread in all populations and also provided novel insights on the level of dissimilarity of individual trophic variation in closely related and ecologically similar newt species [17].

Despite the diverse topics that were discussed by all these papers, other interesting issues involving the role of salamanders in the trophic web and their complex behaviors remain to be elucidated and deserve further attention. For example, aposematic displays and deimatic behaviors of salamanders (i.e., startling visual or auditive signals that distract a predator, giving the attacked prey an opportunity to escape) have received little attention by behavioral ecologists [18]. However, separating aposematism from deimatism in brightly colored salamanders or newts may be challenging [19]. This because the same visual signal

1

may be perceived in a completely different way by animals possessing different visual systems, and indeed, salamanders may be attacked by many different predators that will perceive differently the colors displayed by their potential prey [5].

Conflicts of Interest: The author declares no conflict of interest.

References

1. Wake, D.B.; Koo, M.S. Primer-Amphibians. *Cur. Biol.* **2011**, *28*, R1221–R1242. [CrossRef]
2. Burton, T.M.; Likens, G.E. Salamander populations and biomass in the Hubbard Brook Experimental Forest, New Hampshire. *Copeia* **1975**, *1975*, 541–546. [CrossRef]
3. Ventura, M.; Tiberti, R.; Buchaca, T.; Buñay, D.; Sabás, I.; Miró, A. Why should we preserve fishless high mountain lakes? In *High Mountain Conservation in a Changing World*; Catalan, J., Ninot, J.M., Aniz, M., Eds.; Springer: Berlin/Heidelberg, Germany, 2017; pp. 181–205.
4. Best, M.L.; Welsh, H.H. The trophic role of a forest salamander: Impacts on invertebrates, leaf litter retention, and the humification process. *Ecosphere* **2014**, *5*, 1–19. [CrossRef]
5. Wells, K.D. *The Ecology and Behavior of Amphibians*; University of Chigago Press: Chcago, IL, USA, 2007.
6. Sánchez-Hernández, J. Reciprocal role of salamanders in aquatic energy flow pathways. *Diversity* **2020**, *12*, 32. [CrossRef]
7. Lunghi, E.; Cianferoni, F.; Ceccolini, F.; Zhao, Y.; Manenti, R.; Corti, C.; Ficetola, G.F.; Mancinelli, G. Same diet, different strategies: Variability of individual feeding habits across three populations of ambrosi's cave salamander (*Hydromantes ambrosii*). *Diversity* **2020**, *12*, 180. [CrossRef]
8. Lunghi, E.; Cianferoni, F.; Ceccolini, F.; Merilli, S.; Zhao, Y.; Manenti, R.; Ficetola, G.F.; Corti, C. Ecological observations on hybrid populations of european plethodontid salamanders, genus *Speleomantes*. *Diversity* **2021**, *13*, 285. [CrossRef]
9. Manenti, R.; Lunghi, E.; Barzaghi, B.; Melotto, A.; Falaschi, M.; Ficetola, G.F. Do salamanders limit the abundance of groundwater invertebrates in subterranean habitats? *Diversity* **2020**, *12*, 161. [CrossRef]
10. Salvidio, S.; Costa, A.; Oneto, F.; Pastorino, M.V. Variability of a subterranean prey-predator community in space and time. *Diversity* **2020**, *12*, 17. [CrossRef]
11. Culver, D.C.; Pipan, T. *Shallow Subterranean Habitats: Ecology, Evolution, and Conservation*; Oxford University Press: Oxford, UK, 2014.
12. Ficetola, G.F.; Canedoli, C.; Stoch, F. The Racovitzan impediment and the hidden biodiversity of unexplored environments. *Cons. Biol.* **2019**, *33*, 214–216. [CrossRef]
13. Salvidio, S.; Romano, A.; Palumbi, G.; Costa, A. Safe caves and dangerous forests? Predation risk may contribute to salamander colonization of subterranean habitats. *Sci. Nat.* **2017**, *104*, 20. [CrossRef] [PubMed]
14. Marques, A.D.M.; Mata, V.A.; Velo Anton, G. COI metabarcoding provides insights into the highly diverse diet of a generalist salamander, *Salamandra salamandra* (Caudata: Salamandridae). *Diversity* **2022**, *14*, 89. [CrossRef]
15. Roner, L.; Costa, A.; Pedrini, P.; Matteucci, G.; Leonardi, S.; Romano, A. A Midsummer night's diet: Snapshot on trophic strategy of the Alpine aalamander, *Salamandra atra*. *Diversity* **2020**, *12*, 202. [CrossRef]
16. Van Valen, L. Morphological variation and width of ecological niche. *Am. Nat.* **1965**, *99*, 377–390. [CrossRef]
17. Mirabasso, J.M.; Bissattini, A.; Bologna, M.A.; Luiselli, L.; Stellati, L.; Vignoli, L. Feeding strategies of co-occurring newt species across different conditions of syntopy: A test of the "within-population niche variation" hypothesis. *Diversity* **2020**, *12*, 181. [CrossRef]
18. Ruxton, G.D.; Sherratt, T.N.; Speed, M.P. *Avoiding Attack. The Evolutionary Ecology of Crypsis, Warning Signals, and Mimicry*; Oxford University Press: Oxford, UK, 2004.
19. Barbieri, G.; Costa, A.; Salvidio, S. Is the northern spectacled salamander *Salamandrina perspicillata* aposematic? A preliminary test with clay models. *Acta Herpetol.* **2021**, *16*, 123–128. [CrossRef]

Review

Reciprocal Role of Salamanders in Aquatic Energy Flow Pathways

Javier Sánchez-Hernández

Área de Biodiversidad y Conservación, Departamento de Biología y Geología, Física y Química Inorgánica, Universidad Rey Juan Carlos, Móstoles, 28933 Madrid, Spain; javier.sanchezh@urjc.es; Tel.: +34-914-888-158

Received: 19 November 2019; Accepted: 15 January 2020; Published: 17 January 2020

Abstract: Many species of salamanders (newts and salamanders per se) have a pivotal role in energy flow pathways as they include individuals functioning as prey, competitors, and predators. Here, I synthesize historic and contemporary research on the reciprocal ecological role of salamanders as predators and prey in aquatic systems. Salamanders are a keystone in ecosystem functioning through a combination of top–down control, energy transfer, nutrient cycling processes, and carbon retention. The aquatic developmental stages of salamanders are able to feed on a wide variety of invertebrate prey captured close to the bottom as well as on small conspecifics (cannibalism) or other sympatric species, but can also consume terrestrial invertebrates on the water surface. This capacity to consume allochthonous resources (terrestrial invertebrates) highlights the key role of salamanders as couplers of terrestrial and aquatic ecosystems (i.e., aquatic–terrestrial linkages). Salamanders are also an important food resource for other vertebrates such as fish, snakes, and mammals, covering the energy demands of these species at higher trophic levels. This study emphasizes the ecological significance of salamanders in aquatic systems as central players in energy flow pathways, enabling energy mobility among trophic levels (i.e., vertical energy flow) and between freshwater and terrestrial habitats (i.e., lateral energy flow).

Keywords: amphibia; energy flow; habitat coupling; predator–prey interactions; top–down control; trophic cascades; trophic ecology; Urodela

---

1. Introduction

Ever since the concept of energy transfer was introduced in the mid-19th century [1,2], much attention has been paid to exploring predator–prey interactions and energy flow pathways to understand ecosystem complexity and functioning e.g., [3–5]. From a theoretical standpoint, every organism is a potential source of energy for successive categories of consumers [2]. In aquatic food webs, energy and nutrients flow from basal to higher trophic levels (i.e., bottom–up flow to consumers), but also energy and nutrients can be transferred from terrestrial to aquatic systems and vice versa [6–8]. The structure and functioning of aquatic food webs and energy flows are mediated through top–down (i.e., consumers/predators can regulate the abundance of their prey at lower trophic levels) and bottom–up (i.e., the availability of resources can limit the abundance of consumers/predators in the trophic level above them) regulations [8,9]. Many species of newts and salamanders (henceforth termed "salamanders") can impact invertebrate populations at successively lower trophic levels (top–down control), driving trophic cascades [10–12], but they also can be an important energy budget for higher trophic levels (i.e., apex predators such as fish, snakes, and mammals) through bottom–up energy flow pathways [3,13]. Thus, it is reasonable to posit that salamanders can display a pivotal role in aquatic energy flow pathways as these species include individuals functioning as prey, competitors, and predators in the aquatic food webs.

The feeding habits of aquatic developmental stages of salamanders are well documented, being able to feed on a wide variety of invertebrate prey captured close to the bottom as well as terrestrial invertebrates on the water surface [14–18]. In general, salamanders have been considered as generalist–opportunist foragers, with remarkable ontogenetic dietary shifts, and their diets reflect the relative abundance of prey in the environment [14,18–20]. In addition, salamanders have been widely used as model organisms to test key ecological questions related to, for example, top–down trophic cascades, nutrient cycling processes, and carbon retention in ecosystems e.g., [3,12,21,22]. However, little attention has been given to exploring the reciprocal role of salamanders (both prey and predators) in aquatic energy flow pathways. Here, I provide a summary review on the ecological role of salamanders as predators and prey in aquatic systems, addressing the knowledge gaps and highlighting promising areas for future research.

2. Salamanders as Predators

2.1. Diet Description

Primary dietary components for salamanders depend upon the literature consulted. For example, Regester et al. [23] concluded that most of the interspecific dietary variations among salamander larvae assemblages are attributed to copepods (Cyclopidae) and three families of aquatic insects (Chironomidae, Chaoboridae, and Culicidae). By contrast, Sánchez-Hernández et al. [18] observed that the diets of different developmental stages of salamanders (larva, juvenile, and adult) primarily consisted of aquatic invertebrates (mainly Plecoptera, Diptera, and Ephemeroptera). Thus, salamanders in aquatic systems feed mainly on invertebrates captured close to the bottom, such as Ostracoda, Cladocera, Copepoda, Chironomidae, *Asellus*, or trichopteran larvae, and can also consume prey items on the water surface [14,15,17,19,24,25]. In addition, some salamanders such as, for example, the Pacific giant salamander (*Dicaptodon tenebrosus*), can consume juvenile fish [26]. The two-toed amphiuma (*Amphiuma means*) was also observed to prey on juvenile green sunfish (*Lepomis cyanellus*) [27]. The consumption of conspecifics (cannibalism) or other sympatric species is often observed in larval salamanders [27–29]. Overall, salamander species can adapt their diets in response to spatial and temporal variation in prey availability [30–32].

Since salamanders are gape-limited predators, their feeding is closely related to prey characteristics, such as prey size [33–35]. Therefore, it is thought that prey items with small size, high body flexibility, without hard external structures (cases/shells), or with weak agility may be easy to capture and handle [36] according to the optimal foraging theory (OFT) [37–39]. Salamanders are able to consume highly mobile (e.g., Baetidae) and less mobile (e.g., Limnephilidae and gastropods) prey [14], although they may also accidentally consume immobile items such as vegetal fragments [40]. Nevertheless, some immobile items, such as eggs, are taken deliberately, and most researchers consider eggs a high-energy source for amphibians [41,42]. Thus, the view of diet description in terms of prey traits can be an important feature to generalize concepts across multiple systems or species. In this regard, prey trait analysis has been proposed as a functional approach to understand mechanisms involved in predator–prey relationships [36,43,44], and consequently, it may be useful for understanding prey-handling efficiency, interspecific interactions, and the mechanisms that determine food partitioning among species [36,45]. It is important to keep in mind that the feeding behavior and prey-handling efficiency of salamanders have been mainly studied in laboratory conditions [46–48], and thus future empirical studies attempting to integrate prey characteristics (e.g., prey traits or prey guild members) with diet descriptions will help to understand feeding and the key role of salamanders as predators.

Salamanders normally undergo a change in diet composition during ontogeny as well as across seasons e.g., [14,16,18]. Metamorphs (i.e., individuals already metamorphosed) rely highly on terrestrial invertebrates, whereas paedomorphs (i.e., individuals with larval traits) prefer planktonic resources [16]. In fact, larval salamanders feed mainly on aquatic invertebrates, whereas adults usually eat more terrestrial invertebrates on the water surface or even forage in terrestrial environments [14,15,30,49,50]. Some species, such as the Pyrenean newt (*Euproctus asper*), can undergo discrete ontogenetic dietary shifts with no major ontogenetic shifts in prey-type consumption but shifts in maximum prey width and terrestrial invertebrate consumption [18]. Most salamander species, on the other hand, exhibit remarkable ontogenetic dietary shifts associated with metamorphosis and concomitant shifts in habitat use (between freshwater and terrestrial environments) [14–16,30,49,50]. Thus, ontogenetic dietary shifts can be explained through shifts in habitat utilization for feeding and body size (i.e., processes scaling with mouth size). First, terrestrial invertebrates are more frequently consumed by metamorphosed individuals, which is linked to their capacity to forage in terrestrial habitats and, second, individuals within a population can share the same prey categories but partitioning of the food resources can occur at prey-size level, in line with limitations imposed by mouth gape.

The diets of salamanders can change across seasons regardless of geographic area (e.g., Europe or America) e.g., [16,51]. Lunghi et al. [52] demonstrated that prey size and richness differed significantly between seasons (in both cases, higher in autumn than in spring). An illustrative example of changes in diet composition across seasons can be found in Vignoli [16], who observed that feeding intensity (number of prey), niche breadth, and diet composition changed noticeably across seasons. More specifically, the dominant prey of paedomorphs changed between summer (planktonic crustaceans) and autumn (terrestrial arthropods); but they remained similar for metamorphs (terrestrial arthropods; Homoptera in summer and aphids in autumn) [16]. Thus, seasonal changes in diet composition can be discrete in some cases, as highlighted for ontogenetic dietary shifts. By contrast, high similarity has been found in the Pyrenean newts, with the same dietary components (Diptera and Plecoptera) dominant in all seasons [18]. However, a distinct use of prey taxa and size was observed over the seasons, as juvenile individuals tended to consume fewer Diptera and Plecoptera in spring and autumn, respectively [18]. One of the main criticisms of the current knowledge of the trophic ecology of salamanders is that many studies have focused on only a single season e.g., [17,19,25], despite ecological processes, such as predator–prey interactions, usually having marked seasonality.

Knowledge of the niche use of salamanders has become pivotal to understanding growth and survival through predation and competition. Although many studies rely on a single model organism, the theory behind niche use needs to be set in a broad ecological framework that includes a delineation of general patterns in the feeding of salamanders. In this regard, this review demonstrates that dietary patterns can be common among salamander species. These patterns include larval individuals preying on aquatic taxa that are easy to capture and handle, such as zooplankton in lentic systems and aquatic insects (Diptera, Ephemeroptera, and Plecoptera) in lotic systems, but there is a greater dependence on terrestrial invertebrates after metamorphosis. As already pointed out, the diet composition of salamanders can be explained by site-specific prey community structures and intrinsic factors related to body size (i.e., processes scaling with mouth size). However, it is still unexplored if the diet composition of salamanders follows broadscale patterns and whether environmental factors related to biogeography can be important for determining global patterns in the feeding of salamander species, as previously demonstrated in other vertebrate taxa [53,54]. This represents a promising avenue for future research that needs to be explored in order to accept or refute the view that the diet composition of salamanders can converge across multiple systems or species.

2.2. Top–Down Control

Salamanders provide direct and indirect biotic control of species diversity and ecosystem processes along grazer and detritus pathways [7]. Salamanders may act as top predators in systems without fish and semi-aquatic vertebrates (e.g., snakes and mammals), such as lentic systems (lakes and ponds) and headwater streams where apex predators are absent e.g., [55–58]. For example, Wissinger et al. [57] highlighted the effects of salamander predation on the invertebrate communities of subalpine wetlands, showing a key influence of salamanders on the composition of benthic and planktonic assemblages. Urban [59] provided a good example about the understanding of the functioning and top–down regulations in salamander assemblages. That is, intermediate consumers (spotted salamanders, *Ambystoma maculatum*) can exacerbate prey biomass declines associated with apex predation (marbled salamander, *Ambystoma opacum*), but buffer the top–down effects of marbled salamander predation on prey diversity [59]. Thus, the strength of top–down effects via salamanders can vary with the structure of the salamander assemblage. In some scenarios, salamander-mediated top–down effects can be weak, because interactions between apex predators and intermediate consumers are very strong (i.e., they select for compensatory evolutionary responses in other species) instead of classical top–down considerations (i.e., apex predators have remarkable effects on lower trophic levels) [59]. Atlas and Palen [60] showed theoretically (using a multitrophic modeling framework) that prey vulnerability (i.e., vulnerable versus armored aquatic invertebrates) can limit the top–down effect of the Pacific giant salamander. That is, the absence of armored aquatic invertebrates (e.g., Order Trichoptera and Coleoptera) exacerbates top–down trophic cascades (reducing insect emergence, but increasing algal and detrital biomass) [60]. This conclusion is supported by the above-mentioned gape limitations and prey-handling efficiency of salamanders [36]; thus, armored prey items with hard external structures (cases/shells) can limit salamander-mediated top–down effects, because those prey items are difficult to capture and handle. However, empirical research under controlled (e.g., mesocosms) or natural conditions is required for this to be accepted as a general theory. The top–down model with salamanders as top predators exemplifies a case where salamander predation is the strongest driver of changes in invertebrate and primary producers (i.e., salamanders suppress aquatic invertebrates, which releases freshwater macrophytes, algae, and detritus from herbivory and detritivory, respectively). These top–down effects of salamanders on ecosystem functions (animal community, nutrient dynamics, and primary production) have also been found in terrestrial systems e.g., [61], which underlines the global importance of salamander species on top–down control. However, it should be kept in mind that some studies have reported no evidence of trophic cascades with larval salamanders [62,63] and weak or inconclusive top–down effects of salamanders on invertebrate communities [64–66]. This highlights that salamanders do not always drive top–down effects, but they are definitively important to understand food-web dynamics through predator–prey interactions.

3. Salamanders as Prey

Salamanders are profitable food resources for aquatic apex predators (e.g., snake, otter, and fish species), representing a key taxon to understand energy and nutrient transfer from invertebrates up to higher trophic levels (Figure 1). Preston and Johnson [67] highlighted the key importance of amphibians as food resources for apex predators, which in turn is key for the understanding of apex predator distributions. Therefore, declines of amphibian species may potentially drive negative consequences for apex predators that prey on amphibians [67]. Jobe et al. [68] recently listed 69 salamander species consumed by 89 predators, with snakes being the most frequently reported predator (35% of predations reported), followed by salamanders (24%) and birds (16%). This list would increase as more research becomes available.

3.1. Consumption of Salamanders in Higher Trophic Levels

Although many salamander species develop defensive strategies (i.e., toxic secretions and cryptic and aposematic colorations) to deter predators, the consumption of salamanders by other vertebrates at higher trophic levels has been reported widely e.g., [69–73]. Salamanders provide food for birds and snakes e.g., [68,69,74]. For example, Preston and Johnson [67] found amphibians in most (93% of frequency of occurrence) of the studied aquatic gartersnakes (*Thamnophis atratus*) in California. Another example of predation of salamanders can be found in Escoriza and Hassine [75], who documented the first record of the viperine snake (*Natrix maura*) preying on the endemic Edough newt (*Pleurodeles poireti*) in northeastern Algeria. Willson and Winne [76] observed that the occurrence of the mole salamander (*Ambystoma talpoideum*) accounted for the majority (90%) of the diet composition of two snake species inhabiting the wetlands of South Carolina. Indeed, several researchers accept the view that aquatic salamanders are optimal prey for aquatic snakes because of their morphology traits (i.e., salamanders show elongated morphology), which enables snakes to eat a large number of salamanders with little effects on crawling speed [77].

The consumption of salamanders by mustelid species has also been reported widely e.g., [72,73,78,79]. Smiroldo et al. [80] recently reviewed the amphibians in the diet of the Eurasian otter (*Lutra lutra*) and, in the case of salamanders, the list included eight species (see Table 2 in the work by Smiroldo et al.). However, Jobe et al. [68] only listed one salamander species (great crested newt *Triturus cristatus*) as prey for Eurasian otter (see Appendix A in the work by Jobe et al.). This calls for a better compilation of information to create lists of predators that are capable (i.e., with published records) of feeding upon salamanders. Clavero et al. [81] noticed important seasonal effects in the consumption of amphibians by Eurasian otters, with the highest prevalence during the spawning periods of most amphibian species (late winter–spring). Novais et al. [82] also noticed that amphibians are the most frequent food resource in spring, but fish are the most frequent food resource in autumn, winter, and summer. Parry et al. [79] found that the palmate newt (*Lissotriton helveticus*) remains in Eurasian otter spraints in all months except for February and March. Cogălniceanu et al. [83] observed that Eurasian otters use temporary ponds with high concentrations of amphibians to consume vulnerable prey such as ribbed newts (*Pleurodeles waltl*). These authors also found that Eurasian otters consumed only the soft organs through an incision in the upper part of the thorax [83]. This is most likely because many salamanders produce tetrodotoxin, which is a powerful neurotoxin and anti-predatory skin secretion [84]. Bringsøe and Nørgaard [73] identified the predation of the great crested newt by the Eurasian otter in Denmark. Therefore, predators (here Eurasian otters) can overcome the defensive strategies (i.e., toxic secretions and cryptic and aposematic colorations) of salamanders. Additionally, several researchers support the notion that amphibian consumption by Eurasian otters is inversely related to fish availability (i.e., fish biomass) [85]. Thus, it is possible that spatial and temporal variations in fish availability determine predation on salamanders by mustelid species.

The consumption of salamanders by fish is also well known [54,86–88], with much research focusing on the effect of fish introductions on amphibian assemblages e.g., [89–92]. Overall, amphibian species richness is significantly lower in aquatic systems where fish have been introduced [93], demonstrating the importance of predatory fish species as consumers of salamanders. However, not all amphibian species show the same response to fish introductions e.g., [91,93]. For example, Orizaola and Braña [91] found no effect of salmonid presence on two widespread anuran species (common toad *Bufo bufo* and common midwife toad *Alytes obstetricans*), but they did find a negative impact on newt species (palmate newt, Alpine newt *Triturus alpestris*, and marbled newt *Triturus Marmoratus*) as well as on an anuran species (European common frog *Rana temporaria*). It should be kept in mind that salamanders usually represent a small portion of fish stomach contents [54,94,95]. Taking the brown trout (*Salmo trutta*) as an example, the primary dietary components are commonly Ephemeroptera, Diptera, Trichoptera, Plecoptera, and surface prey (terrestrial arthropods and emerged aquatic insects), with salamander remains recorded only occasionally [54].

The above examples illustrate the importance of salamanders as prey for many vertebrate species. Thus, salamanders are central players in energy and nutrient flow pathways from invertebrates up to higher trophic levels along multiple systems. Knowing the potential consumers of salamanders is integral to understanding the structure and function of communities and ecosystems through predator–prey interactions. This is key to improving our ability to predict the effects of animal introductions or changes in species distributions due to climate change on ecological processes that function at the individual, population, and community levels. From another point of view, Sánchez-Hernández et al. [54] observed an inverse association between mean annual temperature and the global dietary contribution of salamanders in brown trout. Adrián and Delibes [96] concluded that the frequency of occurrence of amphibians in the feces of Eurasian otters tends to decrease with increasing latitude. By contrast, Clavero et al. [53] observed only small differences between Mediterranean and temperate locations in the consumption of amphibians by Eurasian otters. This is seemingly a contradictory standpoint regarding studies in favor of geographical variation in the consumption of salamanders. It is possible that predation upon salamanders follows geographical patterns, representing a particularly promising area for future research. In this regard, the direction of future research could be toward the link between geographical variations and classical observations of latitudinal diversity gradients in amphibians [97,98].

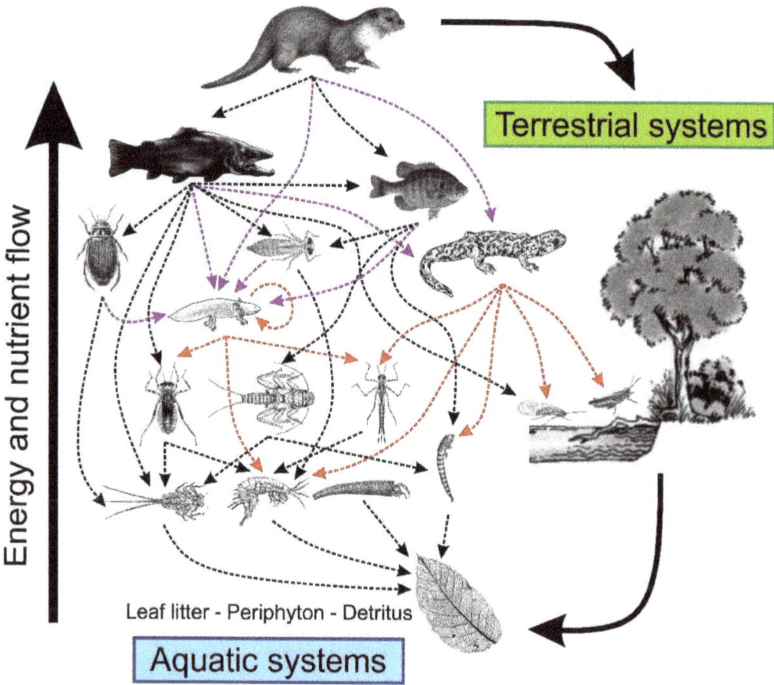

Figure 1. Conceptual view of a hypothetical aquatic food web dominated by semi-aquatic vertebrates (here Eurasian otter, *Lutra lutra*), with salamanders as mid-level vertebrate predators and preys. Red dashed lines represent the interaction of salamanders as predators. Purple dashed lines represent the interaction of salamanders as prey. Arrows indicate the direction of predator–prey interaction and energy flow.

Cannibalism and intraguild predation can also have important implications for the population dynamics, community structure, and distribution patterns of salamanders via growth/mortality trade-offs [99–102]. For example, Resetarits [103] found that the spring salamander (*Gyrinophilus porphyriticus*), a relatively big species, can impact the growth of a smaller salamander (the northern two-lined salamander, *Eurycea bislineata*) and a crayfish (*Cambarus barton*), but have no effect on the fitness (relative condition or fecundity) of brook trout (*Salvelinus fontinalis*). From a broader perspective, the study also recognized the notion that not all salamander species respond in the same way to the presence of predators; some are more vulnerable than those that can behaviorally avoid predation with high growth costs [103]. The consumption of conspecifics (cannibalism) is often observed in larval salamanders [28,29] and needs to be taken into account for a complete understanding of the dynamics and structural properties of populations and food webs. Cannibalism is commonly promoted by low food availability, high conspecific density, and long breeding seasons [104,105]. Long breeding seasons enable interactions among conspecifics of different sizes [105], with bigger individuals (cannibals show proportionally broader heads) being responsible for the cannibalism [104]. Thus, body size emerges as a key precursor of cannibalism, but it is not entirely clear whether body size is a cause (e.g., early hatching and thus bigger individuals having an advantage over smaller or late hatching individuals) or a consequence (e.g., increases in body size are facilitated by cannibalism) of cannibalism [104]. In this regard, the main advantages of cannibalism are the increasing survival and growth rate of individuals that become cannibalistic [106]. Cannibalism and intraguild predation have a key importance to population and community structure via larval production and metamorph emergence, which, in turn, have consequences in energy flow within aquatic systems and between systems (aquatic–terrestrial coupling) [106,107]. These studies demonstrated that cannibalism (intraspecific predation) and intraguild predation represent an energy loop that maintains nutrients and energy within salamander populations or communities, but with high costs because of larval mortality and thus loss of recruitment. In addition, competitive interactions among sympatric salamander species with similar size and structure may reduce the ability to predict changes in both population and community structure [102]. Thus, cannibalism and intraguild predation should not be neglected in studies attempting to explore the influence of population structure, behavior, life history, resource competition, and energy flow. In this regard, cannibalism and intraguild predation have direct consequences on the trophic levels of individuals by feeding on conspecifics and sympatric salamander species, and thus on the energy and nutrient flows in the aquatic ecosystem.

3.2. Salamanders as Energy Subsides for Higher Trophic Levels

Salamanders represent a large standing stock of nutrients in aquatic systems [108], but the nutrient composition of salamander species may change across ontogenetic stages [109]. Although salamanders usually have a lower capacity of nutrient recycling compared to other aquatic vertebrates such as fish species because excretion rates are higher in fish species [110,111], it is expected that salamanders can be important recyclers of biologically essential nutrients in fishless systems [21,111].

Despite there being good examples showing energy pathways in aquatic systems (e.g., leaf litter decomposition and detritus through bacteria into shredder and grazer invertebrates and then directly, or indirectly via mid-level consumers, to the main predators) [112,113], the role of salamanders in energy transfer to higher trophic levels is not completely known. It is thought that salamanders show a highly efficient conversion of invertebrate to vertebrate biomass, but low rates of energy flow compared to birds and mammals [114]. Regester et al. [23] yielded results consistent with the view that assimilation efficiencies for larval marbled salamanders are relatively high, demonstrating that this species efficiently converts ingested prey into biomass. Aquatic salamander communities provide a considerable energy flow (average net flux: 349.5 ± 140.8 g ash-free dry mass per year) into pond habitats [107]. Considering a density of 171 aquatic snakes ha^{-1}, Willson and Winne [76] estimated that aquatic snakes consume over 200 kg (>55,000 individuals) of amphibian prey per year, which represents >150,000 kJ ha^{-1} of energy flow from salamanders (secondary consumers) to snakes (tertiary consumers). Thus, most of our current understanding of energy transfer in aquatic systems provides compelling evidence that salamanders are important energy subsides for higher trophic levels [76,79,108]. However, it is possible that salamanders are largely consumed by predators when salamanders are accessible and vulnerable to predators (i.e., breeding season) or when prey taxa other than salamanders (e.g., fish) are scarce in the environment e.g., [7,85]. Therefore, salamanders have a key role as energy subsides for higher trophic levels, but also need to be considered as a resource pulse (i.e., "episodes of increased resource availability in space and time that combine low frequency, large magnitude, and short duration", which is a definition provided by Yang et al. [115]). Consistent with this view, there is ubiquitous evidence that salamanders can be an important energy source for consumers at higher trophic levels, but often they are spatially and temporally variable.

A better understanding of the role of salamanders in energy transfer demands a complete knowledge of energy inputs and outputs. Research in resource inputs for salamanders has identified the most important prey, in terms of energy inputs, for salamander species. For example, larval ambystomatid salamanders (marbled salamander and spotted salamanders) acquire most of their energy from copepods and larval dipterans [23]. It is possible that freshwater crustaceans other than copepods, such as isopods, cladocerans, and amphipods, may also be a major energy food source for salamanders in aquatic systems [116]. Denoël [117] quantified the origin (aquatic prey versus terrestrial invertebrates) of the energy intake by juvenile Alpine newts, concluding that 62% of the energy intake came from aquatic prey and the other 38% came from terrestrial invertebrates caught in water. Considering that around 10% of the energy ingested during larval development is associated with the production of metamorphosed salamanders [23], it can be expected that salamander species provide a key energy subsidy for higher trophic levels in aquatic systems. For example, newts represent a profitable prey for Eurasian otters [79]. Pagacz and Witczuk [118] estimated that the consumption of amphibians by Eurasian otters can be equally important as fish, constituting 54% of the biomass of consumed prey. Smiroldo et al. [80] revealed that the frequency of occurrence of amphibians in dietary studies in Eurasian otters averaged 12%. Considering that the frequency of occurrence only covers the total number of stomachs with prey and thus without taking the relative abundance of each prey type into account [119], it is doubtful that the use of literature sources only including frequency of occurrence can draw conclusions regarding the importance of salamanders as energy resources for otter species compared to papers reporting the abundance (numerical, biomass, or volume) of prey categories. Salamanders commonly represent a small portion of the stomach contents in brown trout and northern pike (*Exos lucius*) [54,77,78], which underscores the small importance of salamanders as energy subsides for piscivorous fish species such as, for example, many salmonid and pike species. On the other hand, salamanders are thought to be important energy sources for centrarchid species [86,87,120].

Except for works focused on aquatic systems [23,107], remarkable advances in the understanding of energy transfer through salamanders came from studies carried out in terrestrial systems. This is the case of the classical work by Burton and Likens [3], who showed that salamander populations in a forest ecosystem only utilized around 0.02% of the net primary productivity. The authors also provided solid evidence that salamander populations are a remarkable energy source for predators (20% of the energy flow through bird and mammal populations) [3]. Semlitsch et al. [13] suggested that salamanders in forest ecosystems may play a greater role in the trophic transfer of energy and nutrients, as salamander densities were 2–4 times greater than the values previously reported by Burton and Likens [3]. However, to the best of my knowledge, the trophic efficiency mediated by salamanders (i.e., the percentage of energy assimilated by salamanders through the consumption of aquatic invertebrates and the percentage of energy available for higher trophic levels) in aquatic systems has yet to be quantified. The relatively sparse literature on this topic suggests that this would be a fruitful area for future research. Such studies will likely reveal whether trophic efficiency mediated by salamanders can be generalized across multiple systems or species.

4. Salamanders as Promoters of Aquatic–Terrestrial Coupling (Lateral Energy Transfers)

Aquatic and terrestrial systems are not isolated, but interconnected [6,121,122]. Salamanders can couple the link between freshwater and terrestrial habitats through seasonal migrations (breeding adults) and emergences (metamorphosing larvae) e.g., [7,23,107]. Regester et al. [23] demonstrated that metamorphosed salamanders export 3–8% of total prey production to adjacent forest. Thus, the production of metamorphosed salamanders is key for predicting the magnitude of energy subsidies transferred from aquatic systems to terrestrial systems. As already pointed out, cannibalism and intraguild predation represent important regulation mechanisms that are responsible for larval production, and thus of emerging metamorphs [106,107]. In this regard, Regester et al. [107] quantified the grams of ash-free dry mass per year exported from ponds to surrounding forest by emerging metamorphs (between 21 and 135.2 g). This highlights the relevance of salamanders on the energy flow between aquatic and terrestrial habitats, which evidences the aquatic–terrestrial coupling promoted by emerging metamorphs. The coupling of aquatic and terrestrial systems can also be promoted by feeding on allochthonous food resources (terrestrial invertebrates). In fact, inputs of terrestrial invertebrates from riparian canopy cover may represent an important food resource for salamanders inhabiting aquatic environments [18,117]. The importance has been identified in multiple ways: in terms of energy subsidy [117], food resource partitioning between sympatric salamander species [18], or ecosystem functioning [6]. For example, Denoël [117] demonstrated that aquatic juveniles of Alpine newts in Drakolimni lake increase their energy intake using allochthonous food resources up to 60%.

Thus, terrestrial invertebrates can provide high-energy gains [117], and they can represent more than 25% of salamanders' diet in terms of biomass during the summer [123]. This represents a first level of aquatic–terrestrial coupling, by which salamanders rely on allochthonous food resources. However, it is reasonable to posit that, at least in the presence of drift feeders such as salmonids, this aquatic–terrestrial coupling is not promoted by salamanders (diets dominated by benthic aquatic invertebrates), as terrestrial subsides are monopolized by fish species [65].

The aquatic–terrestrial coupling can also be promoted when salamanders act as prey for terrestrial predators ([68] and references therein). For example, it is well known that predatory birds frequently feed on salamander species [68,69], which represents an important link between terrestrial and aquatic habitats. Another level of interface between aquatic and terrestrial habitats can be related to the direct movement of salamanders between these habitats. Due to the life cycle and water dependency of salamanders, they commonly migrate and disperse from their overwintering terrestrial spots to aquatic breeding sites and vice versa [124]. These movements link the energy and nutrient budgets between environments and drive new interactions in the colonized habitat [7,125]. Thus, salamander species intervene in ecosystem processes, enhancing the flux of energy and nutrients between terrestrial and aquatic habitats.

Overall, it is reasonable to posit that aquatic and terrestrial ecosystems can be strongly linked through salamanders, which mobilize nutrients and energy between both interconnected ecosystems (Figure 1). More specifically, this review provides fundamental evidence that the aquatic–terrestrial coupling can be promoted by salamanders at three levels: (i) the consumption of allochthonous food resources by salamanders (i.e., transfer of terrestrial resources to aquatic consumers), (ii) consumption of salamanders by terrestrial predators (i.e., transfer of aquatic resources to terrestrial consumers), and (iii) direct movement of salamanders between aquatic and terrestrial habitats (e.g., seasonal migrations of breeding adults and emergences of metamorphosed larval). These three points where coupling can be mediated by salamander species exemplify the importance of these taxa for understanding unidirectional and bidirectional lateral energy transfers across aquatic and terrestrial ecosystems.

5. Conclusions

This review synthetizes the role of salamanders in aquatic energy flow pathways, pointing out their crucial role in ecosystem functioning as promoters of vertical and lateral energy flows. The role of salamanders in the vertical dimension accepts the view that salamanders promote energy flow among trophic levels, mobilizing energy and nutrients from aquatic invertebrates up to salamanders, but also efficiently converting ingested prey into tissue biomass to become a profitable resource for higher trophic levels (Figure 1). The lateral dimension represents the processes by which energy and nutrients are transferred between aquatic and terrestrial habitats, as well as within trophic levels promoted by cannibalism and intraguild predation in aquatic ecosystems.

Despite research progress on the trophic ecology of salamanders, our knowledge and understanding mostly come from local geographic zones and single model organisms; thus, future research needs to be set in a broad ecological framework to answer global research questions. Studies focusing on the role of salamanders in energy flow pathways will need to be extended to the whole life cycle and ecosystem (aquatic and terrestrial) dimensions to be accepted as a general theory, representing a particularly promising area for future research. An additional promising avenue could be the exploration of global patterns in the trophic ecology of salamanders. In this regard, because high-level salamander phylogenies reflecting the link between diet and morphology (tooth shapes and dentition patterns) are available [126], further studies that integrate geographical and evolutionary frameworks into the feeding and dietary specialization of salamanders are required (see Rabosky et al. [127] for an example using fish as model organisms) to identify the basis of global patterns. Another fruitful area for future research could be exploration of the trophic efficiency of salamanders between trophic levels (i.e., between aquatic invertebrates and apex predators via salamanders as mid-level vertebrate predators). Future studies will likely reveal whether the role of salamanders in energy flow pathways varies geographically along ecosystem (aquatic and terrestrial) or latitudinal dimensions, generating novel insights into the implications of salamanders for communities and ecosystem processes and functioning at global scales.

To date, there is both empirical and theoretical evidence showing that salamanders can be key in aquatic ecosystem functioning through several key processes related to top–down control, becoming an energy source for higher trophic levels, and aquatic–terrestrial coupling (Figure 2). As a caveat, caution should be exercised regarding this conclusion because this figure (i.e., summarizing the number of papers supporting or refuting the potential importance of salamanders in the different roles: aquatic top–down trophic cascades and vertical and lateral energy flows) is based on the literature used in each section (Sections 2.2, 3.2 and 4, respectively) and not on an extensive literature review on the topics. Yet, this review emphasizes the ecological significance of salamanders in aquatic systems as central players in energy flow pathways, enabling energy mobility among trophic levels (i.e., vertical energy flow) and between freshwater and terrestrial habitats (i.e., lateral energy flow).

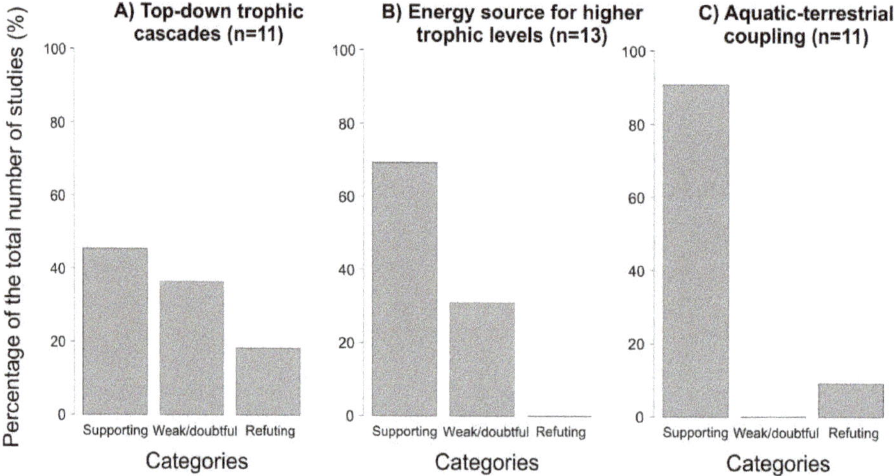

Figure 2. Number of papers supporting or refuting the potential importance of salamanders in aquatic top–down trophic cascades (**A**), energy source for higher trophic levels (i.e., vertical energy flow) (**B**), and energy mobility between freshwater and terrestrial habitats (i.e., lateral energy flow) (**C**). This plot is made according to literature used in each section (Sections 2.2, 3.2 and 4, respectively) of this review. Papers were classified according to three categories: (i) supporting, (ii) weak/doubtful, and (iii) refuting.

Funding: This research received no external funding.

Acknowledgments: Andy Nunn is acknowledged for valuable comments and corrections on the manuscript. I also appreciate constructive comments from three anonymous reviewers, which considerably improved the quality of the manuscript.

Conflicts of Interest: The author declares no conflict of interest.

References

1. Ivlev, V.S. Transformation of energy by aquatic animals. *Int. Rev. Ges. Hydrobiol. Hydrogr.* **1939**, *38*, 449–458. [CrossRef]
2. Lindeman, R. The trophic-dynamic aspect of ecology. *Ecology* **1942**, *23*, 399–418. [CrossRef]
3. Burton, T.M.; Likens, G.E. Energy flow and nutrient cycling in salamander populations in the Hubbard Brook experimental forest, New Hampshire. *Ecology* **1975**, *56*, 1068–1080. [CrossRef]

4. Thompson, R.M.; Brose, U.; Dunne, J.A.; Hall, R.O.; Hladyz, S.; Kitching, R.L.; Martinez, N.D.; Rantala, H.; Romanuk, T.N.; Stouffer, D.B.; et al. Food webs: Reconciling the structure and function of biodiversity. *Trends Ecol. Evol.* **2012**, *27*, 689–697. [CrossRef]

5. Murphy, E.J.; Cavanagh, R.D.; Drinkwater, K.F.; Grant, S.M.; Heymans, J.J.; Hofmann, E.E.; Hunt, G.L.; Johnston, N.M. Understanding the structure and functioning of polar pelagic ecosystems to predict the impacts of change. *Proc. R. Soc. B* **2016**, *283*, 20161646. [CrossRef]

6. Nakano, S.; Miyasaka, H.; Kuhara, N. Terrestrial aquatic linkages: Riparian arthropod inputs alter trophic cascades in a stream food web. *Ecology* **1999**, *80*, 2435–2441.

7. Davic, R.D.; Welsh, H.H.J. On the ecological roles of salamanders. *Annu. Rev. Ecol. Evol. Syst.* **2004**, *35*, 405–434. [CrossRef]

8. Power, M.E.; Holomuzki, J.R.; Lowe, R.L. Food webs in Mediterranean rivers. *Hydrobiologia* **2013**, *719*, 119–136. [CrossRef]

9. Allan, J.D.; Castillo, M.M. *Stream Ecology Structure and Function of Running Waters*, 2nd ed.; Chapman and Hall: New York, NY, USA, 2007.

10. Holomuzki, J.R.; Collins, J.P.; Brunkow, P.E. Trophic control of fishless ponds by tiger salamander larvae. *Oikos* **1994**, *71*, 55–64. [CrossRef]

11. Hocking, D.J.; Babbitt, K.J. Effects of Red-Backed Salamanders on Ecosystem Functions. *PLoS ONE* **2014**, *9*, e86854. [CrossRef]

12. Nery, T.; Schmera, D. The effects of top-down and bottom-up controls on macroinvertebrate assemblages in headwater streams. *Hydrobiologia* **2016**, *763*, 173–181. [CrossRef]

13. Semlitsch, R.D.; O'Donnell, K.M.; Thompson, F.R. Abundance, biomass production, nutrient content, and the possible role of terrestrial salamanders in Missouri Ozark forest ecosystems. *Can. J. Zool.* **2014**, *92*, 997–1004. [CrossRef]

14. Fasola, M.; Canova, L. Feeding habits of *Triturus vulgaris*, *T. cristatus* and *T. alpestris* (Amphibia, Urodela) in the northern Apennines (Italy). *Boll Zool.* **1992**, *59*, 273–280. [CrossRef]

15. Montori, A. Alimentación de las larvas de tritón pirenaico, *Euproctus asper*, en el prepirineo de la Cerdaña, España. *Amphibia–Reptilia* **1992**, *13*, 157–167. [CrossRef]

16. Vignoli, L.; Bombi, P.; D'Amen, M.; Bologna, M.A. Seasonal variation in the trophic niche of a heterochronic population of *Triturus alpestris apuanus* from the south-western Alps. *Herpetol. J.* **2007**, *17*, 183–191.

17. Romano, A.; Salvidio, S.; Palozzi, R.; Sbordoni, V. Diet of the newt, *Triturus carnifex* (Laurenti, 1768), in the flooded karst sinkhole Pozzo del Merro, central Italy. *J. Caves Karst. Stud.* **2012**, *74*, 271–277. [CrossRef]

18. Sánchez-Hernández, J.; Montori, A.; Llorente, G.A. Ontogenetic dietary shifts and food resource partitioning in a stream-dwelling Urodela community: Mechanisms to allow coexistence across seasons. *Russ. J. Herpetol.* **2019**, *26*, 135–149. [CrossRef]

19. Joly, P.; Giacoma, C. Limitation of similarity and feeding habits in three syntopic species of newts (*Triturus*, Amphibia). *Ecography* **1992**, *15*, 401–411. [CrossRef]

20. Salvidio, S.; Oneto, F.; Ottonello, D.; Costa, A.; Romano, A. Trophic specialization at the individual level in a terrestrial generalist salamander. *Can. J. Zool.* **2015**, *93*, 79–83. [CrossRef]

21. Keitzer, S.C.; Goforthm, R.R. Spatial and seasonal variation in the ecological significance of nutrient recycling by larval salamanders in Appalachian headwater streams. *Freshw. Sci.* **2013**, *32*, 1136–1147. [CrossRef]

22. Best, M.L.; Welsh, H.H. The trophic role of a forest salamander: Impacts on invertebrates, leaf litter retention, and the humification process. *Ecosphere* **2014**, *5*, 16. [CrossRef]

23. Regester, K.J.; Whiles, M.R.; Lips, K.R. Variation in the trophic basis of production and energy flow associated with emergence of larval salamander assemblages from forest ponds. *Freshwat. Biol.* **2008**, *53*, 1754–1767. [CrossRef]

24. Braña, F.; de la Hoz, M.; Lastra, C. Alimentación y relaciones tróficas entre las larvas de *Triturus marmoratus*, *T. alpestris* y *T. helveticus* (Amphibia, Caudata). *Doñana Acta Vert.* **1986**, *13*, 21–33.

25. Rosa, G.; Costa, A.; Salvidio, S. Trophic strategies of two newt populations living in contrasting habitats on a Mediterranean island. *Ethol. Ecol. Evol.* **2019**. [CrossRef]

26. Parker, M.S. Predation by Pacific giant salamander larvae on juvenile steelhead trout. *Northwest Nat.* **1993**, *74*, 77–81. [CrossRef]

27. Montaña, C.G.; Ceneviva-Bastos, M.; Schalk, C.M. New vertebrate prey for the aquatic salamander *Amphiuma means* (Caudata: Amphiumidae). *Herpetol. Notes* **2014**, *7*, 755–756.

28. Park, S.-R.; Jeong, J.-Y.; Park, D. Cannibalism in the Korean salamander (*Hynobius leechii*: Hynobiidae, caudata, amphibia) larvae. *Integr. Biosci.* **2005**, *9*, 13–18. [CrossRef]
29. Jefferson, D.M.; Ferrari, M.C.O.; Mathis, A.; Hobson, K.A.; Britzke, E.R.; Crane, A.L.; Blaustein, A.R.; Chivers, D.P. Shifty salamanders: Transient trophic polymorphism and cannibalism within natural populations of larval ambystomatid salamanders. *Front. Zool.* **2014**, *11*, 76. [CrossRef]
30. Guerrero, F.; Perez-Mellado, V.; Gil, M.J.; Lizana, M. Food habits and trophic availability in the high mountain population of the spotted salamander from Spain (*Salamandra salamandra almanzoris*) (Caudata: Salamandridae). *Folia Zool.* **1990**, *89*, 841–853.
31. Jaeger, R.G. Diet diversity and clutch size of aquatic and terrestrial salamanders. *Oecologia* **1981**, *48*, 190–193. [CrossRef]
32. Salvidio, S.; Crovetto, F.; Costa, A. Individual trophic specialisation in the Alpine newt increases with increasing resource diversity. *Ann. Zool. Fenn.* **2019**, *56*, 17–24. [CrossRef]
33. Ranta, E.; Tjossem, S.F.; Leikola, N. Female-male activity and zooplankton foraging by the smooth newt (*Triturus vulgaris*). *Ann. Zool. Fennici.* **1987**, *24*, 79–88.
34. Denoël, M.; Joly, P. Size-related predation reduces intramorph competition in paedomorphic alpine newts. *Can. J. Zool.* **2001**, *79*, 943–948. [CrossRef]
35. Maerz, J.C.; Myers, E.M.; Adams, D.C. Trophic polymorphism in a terrestrial salamander. *Evol. Ecol. Res.* **2006**, *8*, 23–35.
36. Sánchez-Hernández, J. Disentangling prey-handling efficiency of larval newts through multivariate prey trait analysis. *J. Nat. Hist.* **2014**, *48*, 1957–1969. [CrossRef]
37. Emlen, J.M. The role of time and energy in food preference. *Am. Nat.* **1966**, *100*, 611–617. [CrossRef]
38. MacArthur, R.H.; Pianka, E.R. On optimal use of a patchy environment. *Am. Nat.* **1966**, *100*, 603–609. [CrossRef]
39. Pyke, G.H.; Pulliam, H.R.; Charnov, E.L. Optimal foraging: A selective review of theory and tests. *Q. Rev. Biol.* **1977**, *52*, 137–154. [CrossRef]
40. Cicort-Lucaciu, A.-Ş.; Ardeleanu, A.; Cupşa, D.; Naghi, N. The trophic spectrum of a *Triturus cristatus* (Laurentus 1768) population from Plopiş Mountains area (Bihor County, Romania). *North-West J. Zool.* **2005**, *1*, 31–39.
41. Crump, M.L. Opportunistic cannibalism by amphibian larvae in temporary aquatic environments. *Am. Nat.* **1983**, *121*, 281–287. [CrossRef]
42. Crossland, M.R. Predation by tadpoles on toxic toad eggs: The effect of tadpole size on predation success and tadpole survival. *J. Herpetol.* **1998**, *32*, 443–446. [CrossRef]
43. de Crespin de Billy, V.; Usseglio-Polatera, P. Traits of brown trout prey in relation to habitat characteristics and benthic invertebrate communities. *J. Fish Biol.* **2002**, *60*, 687–714. [CrossRef]
44. de Crespin de Billy, V.; Dumont, B.; Lagarrigue, T.; Baran, P.; Statzner, B. Invertebrate accessibility and vulnerability in the analysis of brown trout (*Salmo trutta* L.) summer habitat suitability. *River Res. Appl.* **2002**, *18*, 533–553. [CrossRef]
45. Sánhez-Hernández, J.; Vieira-Lanero, R.; Servia, M.J.; Cobo, F. Feeding habits of four sympatric fish species in the Iberian Peninsula: Keys to understanding coexistence using prey traits. *Hydrobiologia* **2011**, *667*, 119–132. [CrossRef]
46. Ranta, E.; Nuutinen, V. Foraging by the smooth newt (*Triturus vulgaris*) on zooplankton: Functional response and diet choice. *J. Anim. Ecol.* **1985**, *54*, 275–293. [CrossRef]
47. Ranta, E.; Saloheimo, M.; Nuutinen, V. Orientation of the smooth newt (*Triturus vulgaris*) foraging on zooplankton. *Ann Zool Fennici.* **1986**, *23*, 281–287.
48. Heiss, E.; Natchev, N.; Gumpenberger, M.; Weissenbacher, A.; Van Wassenbergh, S. Biomechanics and hydrodynamics of prey capture in the Chinese giant salamander reveal a high-performance jaw-powered suction feeding mechanism. *J. R. Soc. Interface* **2013**, *10*, 20121028. [CrossRef]
49. Kopecký, O.; Vojar, J.; Šusta, F.; Rehák, I. Composing and scaling of male and female alpine newt (*Mesotriton alpestris*) prey, with related site and seasonal effect. *Ann. Zool. Fenn.* **2012**, *9*, 231–239. [CrossRef]
50. Farasat, H.; Sharifi, M. Food habit of the endangered yellow-spotted newt *Neurergus microspilotus* (Caudata, Salamandridae) in Kavat Stream, western Iran. *Zool. Stud.* **2014**, *53*, 61. [CrossRef]

51. Falke, L.P.; Henderson, J.S.; Novak, M.; Preston, D.L. Temporal shifts in intraspecific and interspecific diet variation: Effects of predator body size and identity across seasons in a stream community. *bioRxiv* **2018**. [CrossRef]

52. Lunghi, E.; Cianferoni, F.; Ceccolini, F.; Veith, M.; Manenti, R.; Mancinelli, G.; Corti, C.; Ficetola, G.F. What shapes the trophic niche of European plethodontid salamanders? *PLoS ONE* **2018**, *13*, e0205672. [CrossRef] [PubMed]

53. Clavero, M.; Prenda, J.; Delibes, M. Trophic diversity of the otter (*Lutra lutra* L.) in temperate and Mediterranean freshwater habitats. *J. Biogeogr.* **2003**, *30*, 761–769. [CrossRef]

54. Sánchez-Hernández, J.; Finstad, A.G.; Arnekleiv, J.V.; Kjærstad, G.; Amundsen, P.-A. Drivers of diet patterns in a globally distributed freshwater fish species. *Can. J. Fish. Aquat. Sci.* **2019**, *76*, 1263–1274. [CrossRef]

55. Holomuzki, J.R.; Collins, J.P. Trophic dynamics of a top predator, *Ambystoma tigrinum nebulosum* (Caudata: Ambystomatidae), in a lentic community. *Copeia* **1987**, *1987*, 949–957. [CrossRef]

56. Parker, M.S. Feeding Ecology of Stream-Dwelling Pacific Giant Salamander Larvae (*Dicamptodon tenebrosus*). *Copeia* **1994**, *1994*, 705–718. [CrossRef]

57. Wissinger, S.A.; Whiteman, H.H.; Sparks, G.B.; Rouse, G.L.; Brown, W.S. Foraging trade-offs along a predator-permanence gradient in subalpine wetlands. *Ecology* **1999**, *80*, 2102–2116.

58. Sánchez-Hernández, J.; Cobo, F.; Amundsen, P.-A. Food Web Topology in High Mountain Lakes. *PLoS ONE* **2015**, *10*, e0143016. [CrossRef]

59. Urban, M.C. Evolution mediates the effects of apex predation on aquatic food webs. *Proc. R. Soc. B* **2013**, *280*, 20130859. [CrossRef]

60. Atlas, W.I.; Palen, W.J. Prey Vulnerability Limits Top-Down Control and Alters Reciprocal Feedbacks in a Subsidized Model Food Web. *PLoS ONE* **2014**, *9*, e85830. [CrossRef]

61. Hickerson, C.A.M.; Anthony, C.D.; Walton, B.M. Eastern Red-backed Salamanders regulate top-down effects in a temperate forest-floor community. *Herpetologica* **2017**, *73*, 180–189. [CrossRef]

62. Anderson, T.L.; Rowland, F.E.; Semlitsch, R.D. Variation in phenology and density differentially affects predator–prey interactions between salamanders. *Oecologia* **2017**, *185*, 475–486. [CrossRef] [PubMed]

63. Rowland, F.E.; Rawlings, M.B.; Semlitsch, R.D. Joint effects of resources and amphibians on pond ecosystems. *Oecologia* **2017**, *183*, 237–247. [CrossRef] [PubMed]

64. Reice, S.R.; Edwards, R.L. The effect of vertebrate predation on lotic macroinvertebrate communities in Quebec, Canada. *Can. J. Zool.* **1986**, *64*, 1930–1936. [CrossRef]

65. Atlas, W.I.; Palen, W.J.; Courcelles, D.M.; Munshaw, R.G.; Monteith, Z.L. Dependence of stream predators on terrestrial prey fluxes: Food web responses to subsidized predation. *Ecosphere* **2013**, *4*, 69. [CrossRef]

66. Reinhardt, T.; Brauns, M.; Steinfartz, S.; Weitere, M. Effects of salamander larvae on food webs in highly subsidised ephemeral ponds. *Hydrobiologia* **2017**, *799*, 37–48. [CrossRef]

67. Preston, D.L.; Johnson, P.T.J. Importance of Native Amphibians in the Diet and Distribution of the Aquatic Gartersnake (*Thamnophis atratus*) in the San Francisco Bay Area of California. *J. Herpetol.* **2012**, *46*, 221–227. [CrossRef]

68. Jobe, K.L.; Montaña, C.G.; Schalk, C.M. Emergent patterns between salamander prey and their predators. *Food Webs* **2019**. [CrossRef]

69. Brandon, R.A.; Huheey, J.E. Diurnal Activity, Avian Predation, and the Question of Warning Coloration and Cryptic Coloration in Salamanders. *Herpetologica* **1975**, *31*, 252–255.

70. Brodie, E.D.; Ridenhour, B.J.; Brodie, E.D. The evolutionary response of predators to dangerous prey: Hotspots and coldspots in the geographic mosaic of coevolution between garter snakes and newts. *Evolution* **2002**, *56*, 2067–2082. [CrossRef]

71. Brodie, E.D.; Feldman, C.R.; Hanifin, C.T.; Motychak, J.E.; Mulcahy, D.G.; Williams, B.L.; Brodie, E.D. Parallel Arms Races between Garter Snakes and Newts Involving Tetrodotoxin as the Phenotypic Interface of Coevolution. *J. Chem. Ecol.* **2005**, *31*, 343–356. [CrossRef]

72. Stokes, A.N.; Ray, A.M.; Buktenica, M.W.; Gall, B.G.; Paulson, E.; Paulson, D.; French, S.S.; Brodie, E.D.; Brodie, E.D. Otter Predation on *Taricha granulosa* and Variation in Tetrodotoxin Levels with Elevation. *Northwest Nat.* **2015**, *96*, 13–21. [CrossRef]

73. Bringsøe, H.; Nørgaard, J. Predation of *Triturus cristatus* (Caudata: Salamandridae) by the Eurasian otter, *Lutra lutra* (Carnivora: Mustelidae). *Herpetol. Notes* **2018**, *11*, 279–280.

74. Gregory, P.T.; Isaac, L.A. Food Habits of the Grass Snake in Southeastern England: Is *Natrix natrix* a Generalist Predator? *J. Herpetol.* **2004**, *38*, 88–95. [CrossRef]

75. Escoriza, D.; Hassine, J.B. First case of predation in *Pleurodeles poireti* (Gervais, 1835). *Bol. Asoc. Herpetol. Esp.* **2017**, *28*, 19–20.

76. Willson, J.D.; Winne, C.T. Evaluating the functional importance of secretive species: A case study of aquatic snake predators in isolated wetlands. *J. Zool.* **2015**, *298*, 266–273. [CrossRef]

77. Willson, J.D.; Hopkins, W.A. Prey morphology constrains the feeding ecology of an aquatic generalist predator. *Ecology* **2011**, *92*, 744–754. [CrossRef]

78. Velo-Antón, G.; Cordero-Rivera, A. Predation by invasive mammals on an insular viviparous population of *Salamandra salamandra*. *Herpetol. Notes* **2011**, *4*, 299–301.

79. Parry, G.S.; Yonow, N.; Forman, D. Predation of newts (Salamandridae, Pleurodelinae) by Eurasian otters *Lutra lutra* (Linnaeus). *Herpetol. Bull.* **2015**, *132*, 9–14.

80. Smiroldo, G.; Villa, A.; Tremolada, P.; Gariano, P.; Balestrieri, A.; Delfino, M. Amphibians in Eurasian otter *Lutra lutra* diet: Osteological identification unveils hidden prey richness and male-biased predation on anurans. *Mammal Rev.* **2019**. [CrossRef]

81. Clavero, M.; Prenda, J.; Delibes, M. Amphibian and reptile consumption by otters (*Lutra lutra*) in a coastal area in Southern Iberian Peninsula. *Herpetol. J.* **2005**, *15*, 125–131.

82. Novais, A.; Sedlmayr, A.; Moreira-Santos, M.; Goncalves, F.; Ribeiro, R. Diet of the otter *Lutra lutra* in an almost pristine Portuguese river: Seasonality and analysis of fish prey through scale and vertebrae keys and length relationships. *Mammalia* **2010**, *74*, 71–81. [CrossRef]

83. Cogălniceanu, D.; Márquez, R.; Beltrán, J.F. Impact of otter (*Lutra lutra*) predation on amphibians in temporary ponds in Southern Spain. *Acta Herpetol.* **2010**, *5*, 217–222.

84. Hanifin, C.T. The chemical and evolutionary ecology of tetrodotoxin (TTX) toxicity in terrestrial vertebrates. *Mar. Drugs* **2010**, *8*, 577–593. [CrossRef] [PubMed]

85. Smiroldo, G.; Gariano, P.; Balestrieri, A.; Manenti, R.; Pini, E.; Tremolada, P. Predation on Amphibians May Enhance Eurasian Otter Recovery in Southern Italy. *Zool. Sci.* **2019**, *36*, 273–283. [CrossRef]

86. Semlitsch, R.D. Interactions between fish and salamander larvae. *Oecologia* **1987**, *72*, 481–486. [CrossRef]

87. Semlitsch, R.D. Allotopic Distribution of Two Salamanders: Effects of Fish Predation and Competitive Interactions. *Copeia* **1988**, *1988*, 290–298. [CrossRef]

88. Domínguez, J.; Pena, J.C. Alimentación del lucio *Esox lucius* en un área de reciente colonización (cuenca del Esla, noroeste de España). Variaciones en función de la talla. *Ecologia* **2001**, *15*, 293–308.

89. Knapp, R.A. Effects of nonnative fish and habitat characteristics on lentic herpetofauna in Yosemite National Park, USA. *Biol. Conserv.* **2005**, *121*, 265–279. [CrossRef]

90. Kats, L.B.; Ferrer, R.P. Alien predators and amphibian declines: Review of two decades of science and the transition to conservation. *Divers. Distrib.* **2003**, *9*, 99–110. [CrossRef]

91. Orizaola, G.; Braña, F. Effect of salmonid introduction and other environmental characteristics on amphibian distribution and abundance in mountain lakes of northern Spain. *Anim. Conservat.* **2006**, *9*, 171–178. [CrossRef]

92. Tiberti, R.; Bogliani, G.; Brighenti, S.; Iacobuzio, R.; Liautaud, K.; Rolla, M.; von Hardenberg, A.; Bassano, B. Recovery of high mountain Alpine lakes after the eradication of introduced brook trout *Salvelinus fontinalis* using non-chemical methods. *Biol. Invasions* **2019**, *21*, 875–894. [CrossRef]

93. Martínez-Solano, I.; Barbadillo, J.; Lapeña, M. Effect of introduced fish on amphibian species richness and densities at a montane assemblage in the Sierra de Neila, Spain. *Herpetol. J.* **2003**, *13*, 167–173.

94. Domínguez, J.; Pena, J.C. Spatio-temporal variation in the diet of northern pike (*Esox lucius*) in a colonised area (Esla basin, NW Spain). *Limnetica* **2000**, *19*, 1–20.

95. Cobo, F.; Sánchez-Hernández, J.; Vieira-Lanero, R.; Servia, M.J. Organic pollution induces domestication-like characteristics in feral populations of brown trout (*Salmo trutta*). *Hydrobiologia* **2013**, *705*, 119–134. [CrossRef]

96. Adrián, M.I.; Delibes, M. Food habits of the otter (*Lutra lutra*) in two habitats of the Doñana National Park, SW Spain. *J. Zool.* **1987**, *212*, 399–406. [CrossRef]

97. Gaston, K.J. Global patterns in biodiversity. *Nature* **2000**, *405*, 220–227. [CrossRef]

98. Willig, M.R.; Kaufman, D.M.; Stevens, R.D. Latitudinal gradients of biodiversity: Pattern, process, scale, and synthesis. *Annu. Rev. Ecol. Evol. Syst.* **2003**, *34*, 273–309. [CrossRef]

99. Gustafson, M.P. Intraguild predation among larval plethodontid salamanders: A field experiment in artificial stream pools. *Oecologia* **1993**, *96*, 271–275. [CrossRef]

100. Yurewicz, K.L. A growth/mortality trade-off in larval salamanders and the coexistence of intraguild predators and prey. *Oecologia* **2004**, *138*, 102–111. [CrossRef]

101. Anderson, T.L.; Semlitsch, R.D. Top predators and habitat complexity alter an intraguild predation module in pond communities. *J. Anim. Ecol.* **2016**, *85*, 548–558. [CrossRef]

102. Anderson, T.L.; Linares, C.; Dodson, K.N.; Semlitsch, R.D. Variability in functional response curves among larval salamanders: Comparisons across species and size classes. *Can. J. Zool.* **2016**, *94*, 23–30. [CrossRef]

103. Resetarits, W.J., Jr. Ecological interactions among predators in experimental stream communities. *Ecology* **1991**, *72*, 1782–1793. [CrossRef]

104. Nyman, S.; Wilkinson, R.F.; Hutcherson, J.E. Cannibalism and size relations in a cohort of larval ringed salamanders (*Ambystoma annulatum*). *J. Herpet.* **1993**, *27*, 78–84. [CrossRef]

105. Wildy, E.L.; Chivers, D.P.; Kiesecker, J.M.; Blaustein, A.R. The effects of food level and conspecific density on biting and cannibalism in larval long-toed salamanders, *Ambystoma macrodactylum*. *Oecologia* **2001**, *128*, 202–209. [CrossRef]

106. Polis, G.A. The Evolution and Dynamics of Intraspecific Predation. *Annu. Rev. Ecol. Evol. Syst.* **1981**, *12*, 225–251. [CrossRef]

107. Regester, K.J.; Lips, K.R.; Whiles, M.R. Energy flow and subsidies associated with the complex life cycle of ambystomatid salamanders in ponds and adjacent forest in southern Illinois. *Oecologia* **2006**, *147*, 303–314. [CrossRef]

108. Milanovich, J.R.; Maerz, J.C.; Rosemond, A.D. Stoichiometry and estimates of nutrient standing stocks of larval salamanders in Appalachian headwater streams. *Freshw. Biol.* **2015**, *60*, 1340–1353. [CrossRef]

109. Prater, C.; Scott, D.E.; Lance, S.L.; Nunziata, S.O.; Sherman, R.; Tomczyk, N.; Capps, K.A.; Jeyasingh, P.D. Understanding variation in salamander ionomes: A nutrient balance approach. *Freshw. Biol.* **2019**, *64*, 294–305. [CrossRef]

110. Munshaw, R.G.; Palen, W.J.; Courcelles, D.M.; Finlay, J.C. Predator-Driven Nutrient Recycling in California Stream Ecosystems. *PLoS ONE* **2013**, *8*, e58542. [CrossRef]

111. Milanovich, J.R.; Hopton, M.E. Stoichiometry of Excreta and Excretion Rates of a Stream-dwelling Plethodontid Salamander. *Copeia* **2016**, *104*, 26–34. [CrossRef]

112. Efford, I.E. Energy transfer in Marion Lake, British Columbia; with particular reference to fish feeding. *Verh. Int. Ver. Limnol.* **1969**, *17*, 104–108. [CrossRef]

113. Wallace, J.B.; Eggert, S.L.; Meyer, J.L.; Webster, J.R. Multiple trophic levels of a forest stream linked to terrestrial litter inputs. *Science* **1997**, *277*, 102–104. [CrossRef]

114. Pough, F.H. Amphibians and reptiles as low energy systems. In *Behavioral Energetics; The Cost of Survival in Vertebrates*; Aspey, W.P., Lustick, S.I., Eds.; Ohio State University Press: Columbus, OH, USA, 1983; pp. 141–188.

115. Yang, L.H.; Bastow, J.L.; Spence, K.O.; Wright, A.N. What can we learn from resource pulses? *Ecology* **2008**, *89*, 621–634. [CrossRef] [PubMed]

116. Whiles, M.R.; Jensen, J.B.; Palis, J.G. Diets of larval flatwoods salamanders, *Ambystoma cingulatum*, from Florida and South Carolina. *J. Herpetol.* **2004**, *38*, 208–214. [CrossRef]

117. Denoël, M. Terrestrial versus aquatic foraging in juvenile Alpine newts (*Triturus alpestris*). *Ecoscience* **2004**, *11*, 404–409. [CrossRef]

118. Pagacz, S.; Witczuk, J. Intensive Exploitation of Amphibians by Eurasian Otter (*Lutra lutra*) in the Wołosaty Stream, Southeastern Poland. *Ann. Zool. Fenn.* **2010**, *47*, 403–410. [CrossRef]

119. Amundsen, P.-A.; Sánchez-Hernández, J. Feeding studies take guts—Critical review and recommendations of methods for stomach contents analysis in fish. *J. Fish Biol.* **2019**, *95*, 1364–1373. [CrossRef]

120. Smith, G.R.; Rettig, J.E.; Mittelbach, G.G.; Valiulis, J.L.; Schaack, S.R. The effects of fish on assemblages of amphibians in ponds: A field experiment. *Freshw. Biol.* **1999**, *41*, 829–837. [CrossRef]

121. Nakano, S.; Murakami, M. Reciprocal subsidies: Dynamic interdependence between terrestrial and aquatic food webs. *Proc. Natl. Acad. Sci. USA* **2001**, *98*, 166–170. [CrossRef]

122. Greene, B.T.; Lowe, W.H.; Likens, G.E. Forest succession and prey availability influence the strength and scale of terrestrial-aquatic linkages in a headwater salamander system. *Freshw. Biol.* **2008**, *53*, 2234–2243. [CrossRef]

123. Montori, A. Alimentación de los adultos de *Euproctus asper* (Dugès, 1852) en la montaña media del Prepirineo catalán (España). *Rev. Esp. Herpetol.* **1991**, *5*, 23–36.
124. Semlitsch, R.D. Differentiating migration and dispersal processes for pond-breeding amphibians. *J. Wildl. Manag.* **2008**, *72*, 260–267. [CrossRef]
125. Reinhardt, T.; Steinfartz, S.; Paetzold, A.; Weitere, M. Linking the evolution of habitat choice to ecosystem functioning: Direct and indirect effects of pond-reproducing fire salamanders on aquatic-terrestrial subsidies. *Oecologia* **2013**, *173*, 281–291. [CrossRef] [PubMed]
126. Gregory, A.L.; Sears, B.R.; Wooten, J.A.; Camp, C.D.; Falk, A.; O'Quin, K.; Pauley, T.K. Evolution of dentition in salamanders: Relative roles of phylogeny and diet. *Biol. J. Linn. Soc.* **2016**, *119*, 960–973. [CrossRef]
127. Rabosky, D.L.; Chang, J.; Title, P.O.; Cowman, P.F.; Sallan, L.; Friedman, M.; Kaschner, K.; Garilao, C.; Near, T.J.; Coll, M.; et al. An inverse latitudinal gradient in speciation rate for marine fishes. *Nature* **2018**, *559*, 392–395. [CrossRef]

Article

COI Metabarcoding Provides Insights into the Highly Diverse Diet of a Generalist Salamander, *Salamandra salamandra* (Caudata: Salamandridae)

Adam J. D. Marques [1,2,3,*], Vanessa A. Mata [1,2] and Guillermo Velo-Antón [1,2,4,*]

1   CIBIO, Centro de Investigação em Biodiversidade e Recursos Genéticos, InBIO Laboratório Associado, Campus de Vairão, Universidade do Porto, 4485-661 Vairão, Portugal; vanessamata@cibio.up.pt
2   BIOPOLIS Program in Genomics, Biodiversity and Land Planning, CIBIO, Campus de Vairão, 4485-661 Vairão, Portugal
3   Departamento de Biologia, Faculdade de Ciências, Universidade do Porto, 4099-002 Porto, Portugal
4   Torre Cacti (Lab 97), Grupo de Ecoloxía Animal, Departamento de Ecoloxía e Bioloxía Animal, Universidade de Vigo, E-36310 Vigo, Spain
*   Correspondence: adam.marques@cibio.up.pt (A.J.D.M.); guillermo.velo@uvigo.es (G.V.-A.)

Abstract: DNA metabarcoding has proven to be an accessible, cost-effective, and non-invasive tool for dietary analysis of predators in situ. Although DNA metabarcoding provides numerous benefits in characterizing diet—such as detecting prey animals that are difficult to visually identify—this method has seen limited application in amphibian species. Here, we used DNA metabarcoding to characterize the diet of fire salamanders (*Salamandra salamandra*) (Linnaeus, 1758) in three distinct regions across the northwestern Iberian Peninsula. To test the efficiency of COI-based metabarcoding in determining salamanders' diet diversity, we compared our COI-based results with results from traditional diet studies from neighboring and distant populations, as well as with recent findings obtained in a DNA metabarcoding study using 18S. Two COI primers were used in combination to investigate the potential impact of primer bias in prey detection. Our COI metabarcoding approach increased taxonomic resolution and supported a generalist diet in *S. salamandra*. Between primers, there were no significant differences in the diversity and richness of prey detected. We observed differences in the prevalence of prey identified between sampling regions both in our study and in other studies of *S. salamandra* diet. This COI metabarcoding study provides recommendations and resources for subsequent research using DNA metabarcoding to study amphibian diets.

Keywords: COI; diet; DNA metabarcoding; prey; salamanders

Citation: Marques, A.J.D.; Mata, V.A.; Velo-Antón, G. COI Metabarcoding Provides Insights into the Highly Diverse Diet of a Generalist Salamander, *Salamandra salamandra* (Caudata: Salamandridae). *Diversity* **2022**, *14*, 89. https://doi.org/10.3390/d14020089

Academic Editor: Sebastiano Salvidio

Received: 31 December 2021
Accepted: 24 January 2022
Published: 28 January 2022

Publisher's Note: MDPI stays neutral with regard to jurisdictional claims in published maps and institutional affiliations.

1. Introduction

Diet studies are fundamental to understanding species' dietary habits [1], food webs [2], and trophic niches [3,4], which are key traits in many ecological processes and for the conservation and management of species and ecosystems. DNA metabarcoding as a means of dietary analysis has been used in many taxonomic groups [5,6], but has been underutilized in amphibians [7]—particularly in salamanders. Salamanders serve an important role as mesofaunal predators [8], often comprising a large portion of ecosystem biomass [9], and have low energy requirements, making them potential energy sinks in ecosystems [10]. Moreover, salamanders may also exert a top-down effect on invertebrate community composition and nutrient cycling [7,11,12], which makes studying the diets of salamander species especially relevant for understanding their role in ecosystem functioning.

Visual inspection of stomach or fecal contents is a useful but inconsistent means of diet characterization in salamanders [13,14]. Stomach contents provide insight into recent consumption [3,15,16], and while stomach-flushing avoids sample mortality, it is an invasive approach to diet analysis [14,16–18]. Fecal content inspection is less invasive, but introduces a bias favoring hard-bodied prey species that are not fully digest [13].

DNA metabarcoding can help to identify prey species that are consumed over a longer period of time, avoids the detection bias against soft-bodied prey, and requires less taxonomic training in prey identification [19]. Similar to visual inspections, DNA metabarcoding can also indicate the preferential diets of salamanders and their role as predators in communities [7,17,20,21], and can inform us about invertebrate biodiversity. To our knowledge, only one study has applied DNA metabarcoding to investigate the diet of adult salamander species. Specifically, Wang et al. [22] used the 18S ribosomal RNA (18S) region to characterize the diets of adult fire salamanders (*Salamandra salamandra*) (Linnaeus, 1758) by collecting fecal samples from three Belgian forests. While 18S has proven useful to detect potential prey items, the use of more informative (i.e., variable) DNA fragments such as the cytochrome c oxidase I (COI) benefits from a large reference database that is supported by the Barcode of Life Data System (BOLD) [23,24], and the relatively high variability of the region allows for high-resolution taxonomic assignment [5,25–27].

This study aims to provide an update on the diet of *S. salamandra* while evaluating the use of COI metabarcoding as an efficient, non-invasive method for diet studies, and comparing it to previous works [15], as well as to gain better insights into the diets and functional roles of salamanders as generalist terrestrial predators [28]. Fecal samples were collected from salamanders across the northwestern Iberian Peninsula. To determine whether a significant difference in prey detection could be attributable to primer bias [29], we compared the performance of two COI primers. Technical considerations include the evaluation of (1) sampling effectiveness to capture prey species, and (2) the usefulness of COI primers as barcodes [7,19,30–33].

2. Materials and Methods

Fecal samples were collected from three distinct regions across the northwestern Iberian Peninsula, including the extended metropolitan area of Porto, a forested area across the Morrazo Peninsula, and the island of Ons, which is mostly dominated by bushes (Figure 1). Nocturnal sampling was conducted in Morrazo and Porto in the spring of 2021, and in Ons during November of 2020, coinciding with the highest annual activity peaks for the species and under suitable climatic conditions (i.e., rainy nights and temperatures of 10–20 °C). Up to 20 individuals from each site were collected and placed in sterilized individual or group containers. All individuals were returned to their original sampling sites.

To mitigate primer bias, two COI-specific primer pairs—fwh1 (fwhF1 5′-YTC HAC WAA YCA YAA RGA YAT YGG-3, fwhR1 5′-ART CAR TTW CCR AAH CCH CC-3′) [34] and LerayXT (jgHCO2198 5′-TAI ACY TCI GGR TGI CCR AAR AAY CA-3′, 185 bp; mlCOIintF-XT 5′-GGW ACW RGW TGR ACW ITI TAY CCY CC-3′; 313 bp) [27,35]—were used. *Salamandra salamandra* COI sequences from NCBI (KX094979 & GQ380404) were used to design blocking primers for fwh1 (Ssal_fwhF1-blk 5′-CAA AGA CAT TGG CAC CCT CTA CCT AAT TTT TGG [SpC3]-3′) and LerayXT (Ssal_mlCO1intF-blk 5′-GAA CAG TCT ACC CCC CCC TTG CCG GAA ATC TGG [SpC3]-3′). Initial PCR mixes comprised 10 μL of Qiagen Multiplex PCR Master Mix, 0.3 μL or 0.4 μL of 10 mM target primer (fwh1 and LerayXT, respectively), 8.0 μL of 10 mM blocking primer, 2 μL of DNA template, and enough water for a final volume of 25 μL [36]. Thermocycling conditions included an initial denaturing step at 95 °C for 15 min, followed by 40 cycles of denaturing at 95 °C for 30 s, annealing at 45 °C for 60 s, extension at 72 °C for 30 s, and a final extension at 72 °C for 10 min. Amplification success was verified by running 2 μL of PCR product on a 2% agarose gel. Successfully amplified PCR product was diluted at 1:3 of the initial concentration, in order to reduce the primer dimer during indexing. Illumina indices were then annealed to the PCR product with a PCR composed of 7 μL of KAPA Taq ReadyMix, 1.4 μL of Nextera index [37], 2.8 μL of DNA template, and enough water for a final volume of 14 μL. Thermocycling conditions followed an initial denaturing step at 95 °C for 15 min, followed by 8 cycles of denaturing at 95 °C for 30 s, annealing at 55 °C for 30 s, extension at 72 °C for 30 s, and a final extension at 72 °C for 10 min. Indexed PCR products were purified with AMPure XP beads (Beckman Coulter), eluted to 25 µL, and pooled into equimolar concentrations per fragment. The pooled libraries were quantified with qPCR, normalized to 4 nM, and sequenced in an Illumina MiSeq with an expected coverage of 20,000 reads per sample.

Figure 1. Maps include the species distribution of *Salamandra salamandra*, our study region in the northwest Iberian Peninsula, and photos illustrating the habitats found within the three sampled regions of Ons, Morrazo, and Porto.

Paired reads were aligned with PEAR [38], and successfully assembled reads went through 'ngsfilter' from OBITools [39] to remove primer sequences and annotate sample information. Trimmed reads were then collapsed into unique sequence variants using 'obiuniq' and denoised with '–cluster-unoise' from VSEARH [40], using default parameters, except for minimum sequence length, which was set as 150 bp for fwh1 and 300 bp for LerayXT. Resulting zero-radius operational taxonomic units (zOTUs) went through chimera removal with '–uchime3_denovo', and were clustered at 99% identity [41] with '–cluster_size'. Finally, reads were mapped to the remaining OTUs with 99% identity using '–usearch_global'. To further remove potential nuclear mitochondrial copies (NUMTs) and surviving PCR and sequencing errors, the R package LULU [42] was used with the default parameters. Extraction and PCR negatives were used to correct for contamination. The maximum number of reads of any OTU identified in either extraction or PCR negative was subtracted from the number of reads observed of that OTU in each sample. OTUs were assigned to a taxon using BOLDIGGER v.1.2.5 [43]. OTUs with a minimum of 90% similarity to a taxon included in the phyla Annelida, Arthropoda, or Mollusca were retained as plausible invertebrate prey [22]. Samples with less than 100 reads assigned to dietary OTUs were discarded, as were OTUs comprising less than 1% of the total dietary reads per sample, so as to avoid errors from tag jumping or overrepresentation of rare prey [44]. Prey items were identified at the genus level, as assignment accuracy at the species level was often missing or undefined in the reference database.

To estimate and compare sampling completeness for each region and fragment, as well as prey richness, we used rarefaction curves based on Hill numbers using the 'iNEXT' function from the iNEXT package in R [45,46]. Prey occurrences for each site and for each fragment were converted into incidence frequencies using 'incfreq', and then sample coverage and prey richness were calculated. Sample coverage gives us the proportion of the diet composed of prey species already sampled, and is considered a better reference than sample size to compare species richness among differently sampled groups [47]. To compare the prey composition among samples of different regions and fragments, we calculated a pairwise distance matrix using the Jaccard dissimilarity indices using 'vegdist' available in the R package vegan [48] to quantify the differences between regions and fragments based on prey occurrence. This matrix was then tested using a permutational multivariate analysis of variance (PERMANOVA) with the Jaccard method and 1000 permutations using 'adonis'. One of the assumptions of PERMANOVA is that there are no differences in dispersion among groups; thus, we further conducted a beta dispersion test using 'betadisper' to confirm this homogeneity. Test results were summarized and displayed via principal coordinates analysis (PCoA). Finally, to assess which prey items were significantly contributing to differences between regions and fragments, we conducted a similarity percentage test using 'simper' with 1000 permutations [49].

3. Results

3.1. Sample Collection and Sequence Amplification

A total of 50 individual fecal pellets were extracted (Morrazo = 32, Porto = 8, Ons = 10), with two replicate extractions from a single pellet collected in Porto, and three extraction negatives. Fragments were successfully amplified in 30 samples with fwh1 (Morrazo = 20, Ons = 10) and 38 samples with LerayXT (Morrazo = 21, Porto = 8, Ons = 9), each including the extraction negatives as well as a PCR negative. Post-filtering, we retained dietary reads from a total of 35 individuals, with 25 samples sequenced at either fragment (83% and 66% success rates for fwh1 and LerayXT, respectively) and 15 samples sequenced at both. Each extraction replicate identified three prey taxa, of which two were common to both replicates, while those prey found in only one replicate comprised less than 5% of the total dietary reads.

3.2. Diet Characterization

Across both fragments, a total of 95 unique OTUs were retained, corresponding to 58 prey taxa (Table 1). Two families of Annelida were identified—Almidae and Lumbricidae—wherein one and five genera were identified, respectively. Included among these, *Lumbricus* (present in 49% of samples) was the most common Annelida prey. Arthropoda was the most diverse phylum, comprising 6 known (1 unknown) classes, 18 orders, 32 known (one unknown) families, and 30 known (9 unknown) genera. Among all of these families, no more than two genera were identified. Arthropoda included some of the most common prey—namely, millipedes (Diplopoda), and in particular the genera *Polydesmus* (present among 51% of all samples), *Glomeris* (31%), and *Ommatoiulus* (26%). Mollusca prey comprised only Gastropoda—namely, the orders Pulmonata and Stylommatophora, the former corresponding to a single genus, *Cochlicella*, and the latter comprising 11 families and 12 genera. The second most common prey overall were roundback slugs of the genus *Arion* (49%). While in some instances species-level resolution was available for some prey (e.g., all OTUs assigned to the genus *Glomeris* were also identified as the species *G. occidentalis*), in many cases, taxonomic assignments were unresolved at the species level (e.g., of the nine OTUs assigned to the genus *Arion*, seven different published sequences were identified as matches, but all lacked species-level designations). To avoid potentially inflating the prey richness as an artifact of unresolved taxonomic assignments, we opted instead to use the genus-level resolution, which was available for the majority of OTUs identified.

Table 1. The frequency of occurrence of prey taxa observed among samples in each region, and in total. Where an OTU at the genus-level resolution could not be identified, the next highest taxonomic resolution (e.g., family, order, etc.) is provided. Significant differences in pairwise comparisons of average abundance between regions are shown in bold with an asterisk ($p < 0.05$).

Phylum	Class	Order	Family	Genus	Morrazo	Ons	Porto	Total
Annelida	Clitellata	Haplotaxida	Almidae	*Alma*		**30 ***		9
			Lumbricidae	*Aporrectodea*		20		6
				Dendrobaena	18			9
				Eisenia	12			6
				Lumbricus	35	30		26
				Octolasion	6			3
Arthropoda					6			3
	Arachnida	Opiliones	Ischyropsalididae	*Ischyropsalis*	12			6
		Trombidiformes	Eupodidae			10		3
	Chilopoda	Scolopendromorpha	Cryptopidae	*Cryptops*	6			3
	Collembola	Entomobryomorpha	Entomobryidae			10		3
			Isotomidae			10		3
		Poduromorpha				10		3
			Hypogastruridae	*Hypogastrura*		**30 ***		9
		Symphypleona	Bourletiellidae			20		6
			Dicyrtomidae	*Dicyrtomina*		10		3
			Sminthurididae			10		3
	Diplopoda	Glomerida	Glomeridae	*Glomeris*	65			31
		Julida	Julidae	*Cylindroiulus*	12			6
				Ommatoiulus	47		13	26
		Platydesmida	Andrognathidae		18			9
		Polydesmida			18			9
			Paradoxosomatidae	*Oxidus*			**25 ***	6
			Polydesmidae	*Polydesmus*	88		38	51
	Insecta	Coleoptera	Cantharidae	*Cantharis*	12			6
			Curculionidae	*Caenopsis*	12			6
			Histeridae	*Pactolinus*		10		3
			Tenebrionidae	*Nalassus*	24			11
		Dermaptera	Forficulidae	*Forficula*		10		3
		Diptera	Dolichopodidae	*Condylostylus*	12			6
			Ephydridae	*Scatella*	12			6
			Psychodidae	*Bichromomyia*		10		3
			Sciaridae	Sciaridae sp.	12			6
			Sepsidae	*Meropliosepsis*	6			3
			Syrphidae	*Eupeodes*			**13 ***	3
		Hemiptera	Aphididae	*Chaitophorus*			**13 ***	3
		Hymenoptera					**25 ***	6
		Lepidoptera	Noctuidae	*Omphaloscelis*		10		3
				Peridroma			**13 ***	3
			Sphingidae	*Manduca*		10		3
		Orthoptera	Tettigoniidae	*Cyrtaspis*	6			3
	Malacostraca	Isopoda	Armadillidiidae	*Armadillidium*		10	**38 ***	11
				Eluma	6		**63 ***	17
			Oniscidae	*Oniscus*	6			3
			Porcellionidae	*Porcellio*	18			9
Mollusca	Gastropoda	Pulmonata	Cochlicellidae	*Cochlicella*		**40 ***		11
		Stylommatophora	Agriolimacidae	*Deroceras*	12	10	13	11
			Arionidae	*Arion*	82	30		49
				Geomalacus	6			3
			Helicidae	*Oestophora*		**40 ***		11
			Hygromiidae	*Portugala*			**25 ***	6
			Lauriidae	*Lauria*		10		3
			Limacidae	*Lehmannia*		20		6
			Milacidae	*Milax*		**40 ***		11
			Geomitridae	*Ponentina*			13	3
			Oxychilidae	*Oxychilus*	12			6
			Testacellidae	*Testacella*		10		3
			Urocyclidae	*Microkerkus*	6			3

3.3. Species Richness

When comparing samples sequenced at both fragments, we observed a near-identical sample coverage of 65% and 62% for fwh1 and LerayXT, respectively, with fwh1 detecting a higher number of prey species (39) than LerayXT (25) (Figure 2a). However, when comparing both fragments at similar levels of sampling completeness, the estimated prey richness did not differ significantly between the two fragments (overlapping 95% confidence intervals of rarefaction curves; Figure 2a). A two-sample *t*-test confirmed that fwh1 produced a higher number of total dietary reads (10,811 ± 9896) post-filtering compared to LerayXT (1802 ± 3015), but no differences in the total number of filtered reads or the ratio of dietary reads to filtered reads ($t = 7.0748$; $p < 0.001$).

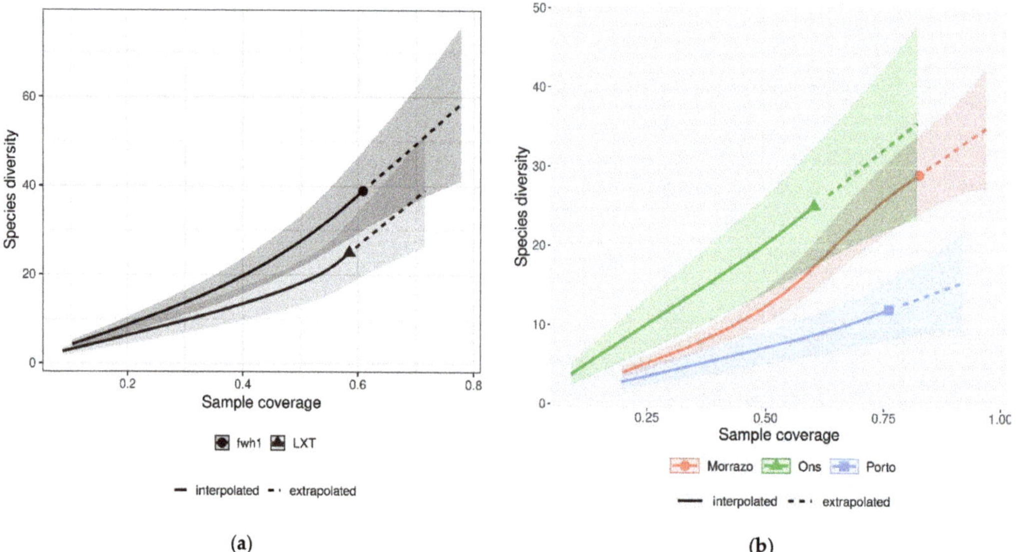

(a) (b)

Figure 2. Species diversity extrapolated as a factor of sample coverage for (**a**) different CO1 fragments; (**b**) sampling regions. Points indicate the observed richness, while error bars indicate the 95% confidence intervals of the response curves.

Sample coverage was 85%, 72%, and 60% for Morrazo, Porto, and Ons, respectively (Figure 2b). Observed prey richness was 29, 25, and 12 for Morrazo, Ons, and Porto, respectively. Rarefaction curves showed that estimated prey richness in Porto was lower than in Morrazo and Ons, while the latter exhibited similar levels of prey richness. The PERMANOVA model showed no differences in the composition of prey identified by either fragment (Figure 3a), but significant differences between regions ($p < 0.001$). However, the beta dispersal test suggests that the significant differences in prey composition observed in the PERMANOVA between sites may be inflated due to the lack of homogeneity in variance across groups (PCoA; Figure 2b; $p = 0.01967$). Notable differences in prey prevalence between regions include the absence of millipedes among samples from Ons. This coincides with an increased diversity of soft-bodied prey from among Ons samples, including several gastropod genera—*Cochlicella*, *Oestophora*, and *Milax*—found to be significantly more common, and the only instance of earthworms from the family *Alma*. Annelida was notably absent among samples from Porto. Several genera of arthropods, however, were significantly more common in Porto than in other locations, including Polydesmida: *Oxidus*, Diptera: *Eupeodes*, Hemiptera: *Chaitophorus*, Lepidoptera: *Peridroma*, and Isopoda: *Armadillidium* and *Eluma*, although we should note that the low sample coverage from Porto may inflate the significance of this observation.

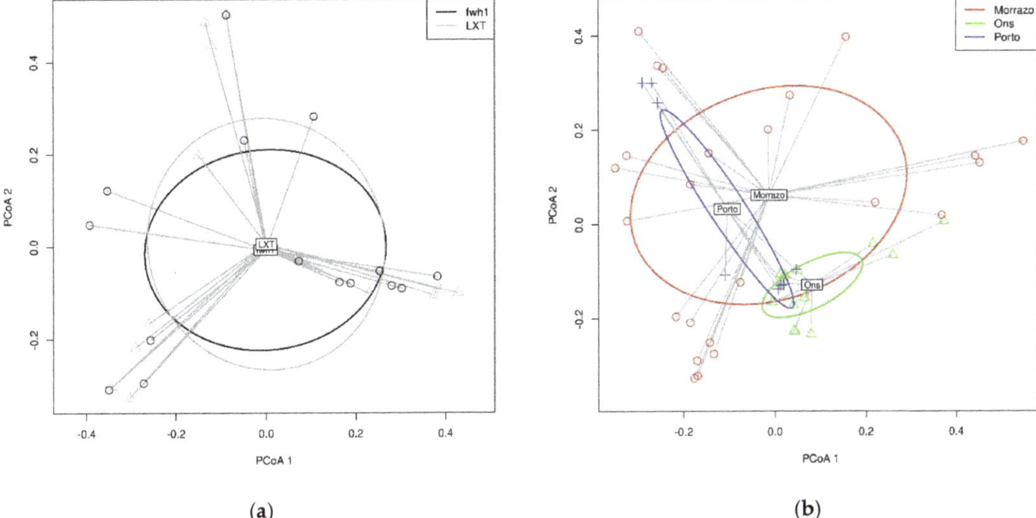

Figure 3. Variance in the prey taxa summarized by the first two eigenvectors of a PCoA for (**a**) different CO1 fragments; (**b**) sampling regions. Ellipses capture the distance of points from the centroid within one standard deviation.

4. Discussion

4.1. COI Metabarcoding for Salamanders' Diet Characterization

Concordant with previous characterizations of *S. salamandra* diet, our results from the DNA metabarcoding of COI suggest a prevalence of low-mobility terrestrial detritivores, including millipedes (Diplopoda), roundback slugs (Arionidae), and earthworms (Lumbricidae) [15–17,50]. While each of these prey taxa are similarly prevalent in the salamander diet as a whole, regional differences such as the absence of Annelids in Porto and Diplopoda in Ons were observed. Comparing the results of COI metabarcoding to diet inspections by Bas et al. [15] in *S. s. gallaica* (northwest Iberia), we observed a clear difference in the types of prey detected. The most common prey identified by Bas et al. [15] were Insecta, with far fewer soft-bodied prey compared to the findings of this study. This was expected, as visual inspection may favor the detection of hard-bodied prey that are more slowly digested compared to soft-bodied prey, which may become visually unrecognizable several days after consumption [13]. We observed this anecdotally in the prevalence of Coleoptera identified by Bas et al. [15], of which 16 genera were identified, as compared to the four genera identified in this study, with both studies identifying *Nalassus* as the most common prey among the class Insecta. Conversely, 18S metabarcoding by Wang et al. [22] found that the most common prey were Gastropoda, which far exceeded other prey in prevalence. Gastropoda have been reported as a common prey elsewhere [16,50], as well as in our results. However, in many of these cases, often only a few prey taxa could be identified, because of either digestion or low taxonomic resolution. Our study identified 58 different prey taxa from three populations at variable taxonomic resolutions—fewer than the 76 prey taxa identified by gastric inspection from a higher number of individuals (N = 72), localities (N = 10), and a wider environmental and ecological survey [15], but nearly threefold more than from previous taxonomic assignment using DNA metabarcoding of 18S across three forest populations in Belgium [22]. Thus, DNA metabarcoding of COI increases taxonomic resolution and provides a cost-effective and expedient method for characterizing the diets of salamanders. Indeed, the increased taxonomic resolution provided by COI suggests that salamanders are able to utilize a variety of Gastropoda prey. This was most evident among our samples from Ons, which often comprised soft-bodied gastropods and a complete absence of Diplopoda—possibly as a result of differences in region, season, or habitat compared to our other sites. For instance, seasonal variation was found to influence prey richness in the diets of salamanders, with the greatest prey richness reported in autumn [4,28]. This seems a probable explanation for the observed prey richness among samples from Ons sampled during autumn. Relatively few arachnid (Arachnida) and centipede (Chilopoda) taxa were identified among our samples, although they have been reported in the diets of salamanders [15,22], suggesting an absence either among our samples or in local abundance.

4.2. COI Primers as Barcodes

While studies should always strive to include variable fragments in order to account for primer biases [29,30], our results suggest no discernible difference in the results gathered by using either *fwh1* or *LerayXT*. Greater species richness among fwh1 sequences compared to LerayXT was unexpected given previous comparisons of COI primer performances in literature [51], although sample degradation may favor the shorter fragment. The absence of any clear differences in the prey composition between these primers, however, casts doubt on whether LerayXT underperformed, as inequalities in read output and sample coverage discourage definitive conclusions. Further sequencing and more distributive sampling will be necessary for verification. The number of dietary reads generated by LerayXT was significantly lower than the number generated by fwh1, even with no discernible difference in the prey being identified by either primer. During sequencing, the smaller fragment—in this case, fwh1—will usually be favored [52]; however, a comparison between COI primers found that fwh1 has a higher likelihood of mismatch between the primer and the template, potentially identifying fewer prey species [51]. Despite expectations, LerayXT identified fewer prey species, with a possible explanation being among the unfiltered reads, as 32% of all OTUs and 26% of all reads were 352 bp—longer than the target fragment length—and either unassigned or identified as *Flavobacterium*. When present, these OTUs were the most abundant reads among a subset of individuals, and may represent an instance of nuclear mitochondrial DNA (NUMT) resulting from transposition of COI into the nuclear genome [53]. Pseudogenes such as these may evade blocking primers and dominate the amplification reaction. No OTUs were assigned to *Salamandra salamandra*, indicating high efficiency of the blocking primers; however, considering the large size and repetitive nature of the salamander genome, pseudogenes are to be expected [54].

4.3. Dietary Variation across Regions

Although differences in prey prevalence were observed between regions, overlap in the prey composition should deter us from drawing any premature conclusions about diet preferences or prey abundance. Instead, we can refer to these preliminary results as a starting point for subsequent studies. Based on our extended rarefaction results (Supplementary Materials Figure S2), we also recommend that future studies aim for a minimum sample size of 20 sample units per site for sample coverage. The differences in prey taxa observed between Ons and the mainland regions—primarily the prevalence of soft-bodied prey such as land snails and slugs (Stylommatophora) and segmented worms (*Alma*) that were otherwise undetected in the mainland samples—is of particular interest. We might have expected islands to have lower alpha diversity than the mainland [55]; however, samples from Ons were temporally distinct from those of Morrazo and Porto, and must be resampled in the same temporal period in order to control for the known effects that seasonal variation has on prey richness [4,19,50]. Additional observations, such as the relative absence of Diplopoda in the diet of Ons samples, may indicate a scarcity of this common prey, driving prey diversification [56]. In Porto, conversely, we anticipated higher prey richness by taxa—namely, Isopoda and Limacidae—able to utilize anthropogenic spaces [57]. Although prey richness was low when compared to other regions, there was a prevalence of pill woodlice (Armadillidiidae), which may be of interest for studies in ecotoxicology, as terrestrial Isopoda often serve as model organisms in soil ecotoxicology [58]. However, we must also consider that these samples were not sequenced at fwh1, due to poor amplification, and have fewer overall reads to compare ($t = 2.1898$; $p < 0.05$). Previous studies investigating prey consumption in *S. salamandra* detected dietary differences between sexes [22], ages [59], seasons [50], and populations [15]. Follow-up studies should also consider comprehensive sampling of distinct habitats throughout the species' range, as well as the remarkable intraspecific differentiation in reproductive modes [60], head shape [61], and behavioral strategies [62–64], both between and within subspecies of *S. salamandra*. The inclusion of these variables may help to elucidate the factors that contribute to the dietary variation observed in this study.

Supplementary Materials: The following are available online at https://www.mdpi.com/article/10.3390/d14020089/s1, Figure S1: Rarefaction comparison between fragments; Figure S2: Rarefaction comparison between regions; Table S1: Extended list of OTUs after sequence annotation and filtration.

Author Contributions: Conceptualization, G.V.-A. and V.A.M.; methodology, V.A.M.; validation, A.J.D.M. and V.A.M.; formal analysis, A.J.D.M.; investigation, G.V.-A. and A.J.D.M.; resources, G.V.-A. and V.A.M.; data curation, G.V.-A., A.J.D.M. and V.A.M.; writing—original draft preparation, A.J.D.M.; writing—review and editing, G.V.-A. and V.A.M.; visualization, A.J.D.M.; supervision, G.V.-A. and V.A.M.; project administration, G.V.-A.; funding acquisition, G.V.-A. All authors have read and agreed to the published version of the manuscript.

Funding: This research was funded by the Fundação para a Ciência e a Tecnologia, grant numbers PTDC/BIA-EVL/28475/2017 and PTDC/BIA-CBI/2278/2020. V.A.M. was funded by the FCT through the program 'Stimulus of Scientific Employment, Individual Support—3rd Edition' (2020.02547.CEECIND). G.V.-A. was supported by the FCT (CEECIND/00937/2018), and recently by a Ramón y Cajal research grant (Ref. RYC-2019-026959-I/AEI/10.13039/501100011033). This work was co-funded by the project NORTE-01-0246-FEDER-000063, supported by Norte Portugal Regional Operational Programme (NORTE2020), under the PORTUGAL 2020 Partnership Agreement, through the European Regional Development Fund (ERDF).

Institutional Review Board Statement: Fieldwork for collecting individuals and fecal samples was carried out with the corresponding permits from the regional and national administrations (Xunta de Galicia, Ref. EB-031/2021; Portugal Ref. 544/2021/CAPT). Sampling procedures were carried out following the Guidelines for Use of Live Amphibians and Reptiles in Field and Laboratory Research, 2nd Edition, revised by the Herpetological Animal Care and Use Committee (HACC) of the American Society of Ichthyologists and Herpetologists, 2004.

Data Availability Statement: The data presented in this study are openly available in Zenodo at https://zenodo.org/badge/latestdoi/429739169 (accessed on 30 December 2021).

Acknowledgments: We thank Lucia Alarcón-Ríos and Iria Pazos for assistance during fieldwork, and the staff of the *Parque Nacional Marítimo-Terrestre das Illas Atlánticas de Galicia* for providing lodging and facilitating the trips to the island.

Conflicts of Interest: The authors declare no conflict of interest.

References

1. Losos, J.B.; Greene, H.W. Ecological and evolutionary implications of diet in monitor lizards. *Biol. J. Linn. Soc.* **1988**, *35*, 379–407. [CrossRef]
2. Svanbäck, R.; Quevedo, M.; Olsson, J.; Eklöv, P. Individuals in food webs: The relationships between trophic position, omnivory and among-individual diet variation. *Oecologia* **2015**, *178*, 103–114. [CrossRef]
3. Lunghi, E.; Manenti, R.; Cianferoni, F.; Ceccolini, F.; Veith, M.; Corti, C.; Ficetola, G.F.; Mancinell, G. Interspecific and inter-population variation in individual diet specialization: Do environmental factors have a role? *Ecology* **2020**, *101*, e03088. [CrossRef]
4. Lunghi, E.; Cianferoni, F.; Ceccolini, F.; Veith, M.; Manenti, R.; Mancinelli, G.; Corti, C.; Ficetola, G.F. What shapes the trophic niche of European plethodontid salamanders? *PLoS ONE* **2018**, *13*, e0205672. [CrossRef]
5. De Sousa, L.L.; Silva, S.M.; Xavier, R. DNA metabarcoding in diet studies: Unveiling ecological aspects in aquatic and terrestrial ecosystems. *Environ. DNA* **2019**, *1*, 199–214. [CrossRef]
6. Ando, H.; Mukai, H.; Komura, T.; Dewi, T.; Ando, M.; Isagi, Y. Methodological trends and perspectives of animal dietary studies by noninvasive fecal DNA metabarcoding. *Environ. DNA* **2020**, *2*, 391–406. [CrossRef]

7.   Ficetola, G.F.; Manenti, R.; Taberlet, P. Environmental DNA and metabarcoding for the study of amphibians and reptiles: Species distribution, the microbiome, and much more. *Amphibia-Reptilia* **2019**, *40*, 129–148. [CrossRef]

8.   Laking, A.E.; Li, Z.; Goossens, E.; Miñarro, M.; Beukema, W.; Lens, L.; Bonte, D.; Verheyen, K.; Pasmans, F.; Martel, A. Salamander loss alters litter decomposition dynamics. *Sci. Total Environ.* **2021**, *776*, 145994. [CrossRef]

9.   Peterman, W.E.; Crawford, J.A.; Semlitsch, R.D. Productivity and significance of headwater streams: Population structure and biomass of the black-bellied salamander (*Desmognathus quadramaculatus*). *Freshw. Biol.* **2007**, *53*, 347–357. [CrossRef]

10.   Davic, R.D.; Welsh Jr, H.H. On the ecological roles of salamanders. *Annu. Rev. Ecol. Evol. Syst.* **2004**, *35*, 405–434. [CrossRef]

11.   Walton, B.M. Salamanders in forest-floor food webs: Environmental heterogeneity affects the strength of top-down effects. *Pedobiologia* **2005**, *49*, 381–393. [CrossRef]

12.   Hocking, D.J.; Babbitt, K.J. Effects of red-backed salamanders on ecosystem functions. *PLoS ONE* **2014**, *9*, e86854. [CrossRef] [PubMed]

13.   Costa, A.; Salvidio, S.; Posillico, M.; Altea, T.; Matteucci, G.; Romano, A. What goes in does not come out: Different non-lethal dietary methods give contradictory interpretations of prey selectivity in amphibians. *Amphibia-Reptilia* **2014**, *35*, 255–262. [CrossRef]

14.   Crovetto, F.; Romano, A.; Salvidio, S. Comparison of two non-lethal methods for dietary studies in terrestrial salamanders. *Wildl. Res.* **2012**, *39*, 266–270. [CrossRef]

15.   Bas Lopez, S.; Gutian Rivera, J.; Castro Lorenzo, A.d.; Sanchez Canals, J.L. Data on the diet of salamander (*Salamandra salamandra* L.) in Galicia. *Bol. Estac. Cent. Ecol.* **1979**, *8*, 73–78.

16.   Çiçek, K.; Koyun, M.; Tok, C.V. Food composition of the Near Eastern Fire Salamander, *Salamandra infraimmaculata* Martens, 1885 (Amphibia: Urodela: Salamandridae) from Eastern Anatolia. *Zool. Middle East* **2017**, *63*, 130–135. [CrossRef]

17.   Solé, M.; Rödder, D. Dietary assessments of adult amphibians. In *Amphibian Ecology and Conservation: A Handbook of Techniques*; Dodd, C.K., Jr., Ed.; Oxford University Press: Oxford, UK, 2009; pp. 167–184.

18.   Salvidio, S.; Pasmans, F.; Bogaerts, S.; Martel, A.; van de Loo, M.; Romano, A. Consistency in trophic strategies between populations of the Sardinian endemic salamander *Speleomantes imperialis*. *Anim. Biol.* **2017**, *67*, 1–16. [CrossRef]

19.   Alberdi, A.; Aizpurua, O.; Bohmann, K.; Gopalakrishnan, S.; Lynggaard, C.; Nielsen, M.; Gilbert, M.T.P. Promises and pitfalls of using high-throughput sequencing for diet analysis. *Mol. Ecol. Resour.* **2019**, *19*, 327–348. [CrossRef]

20.   Unger, S.D.; Williams, L.A.; Diaz, L.; Jachowski, C.B. DNA barcoding to assess diet of larval eastern hellbenders in North Carolina. *Food Webs* **2020**, *22*, e00134. [CrossRef]

21.   Unger, S.D. Adult female Eastern hellbender *Cryptobranchus alleganiensis* (Cryptobranchidae) conspecific cannibalism confirmed via DNA barcoding. *Herpetol. Notes* **2020**, *13*, 169–170.

22.   Wang, Y.; Smith, H.K.; Goossens, E.; Hertzog, L.; Bletz, M.C.; Bonte, D.; Verheyen, K.; Lens, L.; Vences, M.; Pasmans, F.; et al. Diet diversity and environment determine the intestinal microbiome and bacterial pathogen load of fire salamanders. *Sci. Rep.* **2021**, *11*, 20493. [CrossRef]

23.   Hebert, P.D.N.; Ratnasingham, S.; de Waard, J.R. Barcoding animal life: Cytochrome c oxidase subunit 1 divergences among closely related species. *Proc. R. Soc. Lond. B* **2003**, *270*, S96–S99. [CrossRef]

24.   Ratnasingham, S.; Herbert, P.D.N. BOLD: The Barcode of Life Data System (http://www.barcodinglife.org). *Mol. Ecol. Notes* **2007**, *7*, 355–364. [CrossRef]

25.   Borrell, Y.J.; Miralles, L.; Do Huu, H.; Mohammed-Geba, K.; Garcia-Vazquez, E. DNA in a bottle—Rapid metabarcoding survey for early alerts of invasive species in ports. *PLoS ONE* **2017**, *12*, e0183347. [CrossRef] [PubMed]

26.   Grey, E.K.; Bernatchez, L.; Cassey, P.; Deiner, K.; Deveney, M.; Howland, K.L.; Lacoursière-Roussel, A.; Leong, S.C.Y.; Li, Y.; Olds, B.; et al. Effects of sampling effort on biodiversity patterns estimated from environmental DNA metabarcoding surveys. *Sci. Rep.* **2018**, *8*, 8843. [CrossRef]

27.   Wangensteen, O.S.; Palacín, C.; Guardiola, M.; Turon, X. DNA metabarcoding of littoral hard-bottom communities: High diversity and database gaps revealed by two molecular markers. *PeerJ* **2018**, *6*, e4705. [CrossRef]

28.   Velo-Antón, G.; Buckley, D. Salamandra común—Salamandra salamandra. In *Enciclopedia Virtual de los Vertebrados Españoles*; Salvador, A., Martínez-Solano, I., Eds.; Museo Nacional de Ciencias Naturales: Madrid, Spain, 2015; Available online: www.vertebradosibericos.org/ (accessed on 22 November 2021).

29.   Pereira, A.; Samlali, M.A.; S'Khifa, A.; Slimani, T.; Harris, D.J. A pilot study on the use of DNA metabarcoding for diet analysis in a montane amphibian population from North Africa. *Afric. J. Herpetol.* **2021**, *70*, 68–74. [CrossRef]

30.   Zinger, L.; Bonin, A.; Alsos, I.G.; Bálint, M.; Bik, H.; Boyer, F.; Chariton, A.A.; Creer, S.; Coissac, E.; Deagle, B.E.; et al. DNA metabarcoding-Need for robust experimental design to draw sound ecological conclusions. *Mol. Ecol.* **2019**, *28*, 1857–1862. [CrossRef] [PubMed]

31.   Clarke, L.J.; Soubrier, J.; Weyrich, L.S.; Cooper, A. Environmental metabarcodes for insects: In silico PCR reveals potential for taxonomic bias. *Mol. Ecol. Resour.* **2014**, *14*, 1160–1170. [CrossRef]

32.   Deagle, B.E.; Jarman, S.N.; Coissac, E.; Pompanon, F.; Taberlet, P. DNA metabarcoding and the cytochrome c oxidase subunit I marker: Not a perfect match. *Biol. Lett.* **2014**, *10*, 20140562. [CrossRef]

33.   da Silva, L.P.; Mata, V.A.; Lopes, P.B.; Lopes, R.J.; Beja, P. High-resolution multi-marker DNA metabarcoding reveals sexual dietary differentiation in a bird with minor dimorphism. *Ecol. Evol.* **2020**, *10*, 10364–10373. [CrossRef]

34. Vamos, E.; Elbrecht, V.; Leese, F. Short COI markers for freshwater macroinvertebrate metabarcoding. *PeerJ Prepr.* **2017**, *1*, e14625. [CrossRef]

35. Geller, J.; Meyer, C.; Parker, M.; Hawk, H. Redesign of PCR primers for mitochondrial cytochrome c oxidase subunit I for marine invertebrates and application in all-taxa biotic surveys. *Mol. Ecol. Resour.* **2013**, *13*, 851–861. [CrossRef]

36. Lopes, R.J.; Pinho, C.J.; Santos, B.; Seguro, M.; Mata, V.A.; Egeter, B.; Vasconcelos, R. Intricate trophic links between threatened vertebrates confined to a small island in the Atlantic Ocean. *Ecol. Evol.* **2019**, *9*, 4994–5002. [CrossRef] [PubMed]

37. Adey, A.; Morrison, H.G.; Asan; Xun, X.; Kitzman, J.O.; Turner, E.H.; Stackhouse, B.; MacKenzie, A.P.; Caruccio, N.C.; Zhang, X.; et al. Rapid, low-input, low-bias construction of shotgun fragment libraries by high-density in vitro transposition. *Genome Biol.* **2010**, *11*, R119. [CrossRef] [PubMed]

38. Zhang, J.; Kobert, K.; Flouri, T.; Stamatakis, A. PEAR: A fast and accurate Illumina Paired-End reAd merger. *Bioinformatics* **2014**, *30*, 614–620. [CrossRef]

39. Boyer, F.; Mercier, C.; Bonin, A.; Le Bras, Y.; Taberlet, P.; Coissac, E. obitools: A unix-inspired software package for DNA metabarcoding. *Mol. Ecol. Resour.* **2016**, *16*, 176–182. [CrossRef] [PubMed]

40. Rognes, T.; Flouri, T.; Nichols, B.; Quince, C.; Mahé, F. VSEARCH: A versatile open source tool for metagenomics. *PeerJ* **2016**, *4*, e2584. [CrossRef] [PubMed]

41. Mata, V.A.; Ferreira, S.; Campos, R.M.; da Silva, L.P.; Veríssimo, J.; Corley, M.F.V.; Beja, P. Efficient assessment of nocturnal flying insect communities by combining automatic light traps and DNA metabarcoding. *Environ. DNA* **2020**, *3*, 398–408. [CrossRef]

42. Frøslev, T.G.; Kjøller, R.; Bruun, H.H.; Ejrnæs, R.; Brunbjerg, A.K.; Pietroni, C.; Hansen, A.J. Algorithm for post-clustering curation of DNA amplicon data yields reliable biodiversity estimates. *Nat. Commun.* **2017**, *8*, 1188. [CrossRef]

43. Buchner, D.; Leese, F. BOLDigger—A Python package to identify and organise sequences with the Barcode of Life Data systems. *Metabarcoding Metagenom.* **2020**, *4*, 19–21. [CrossRef]

44. Deagle, B.E.; Thomas, A.C.; McInnes, J.C.; Clarke, L.J.; Vesterinen, E.J.; Clare, E.L.; Kartzinel, T.R.; Paige, E. Counting with DNA in metabarcoding studies: How should we convert sequence reads to dietary data? *Mol. Ecol.* **2019**, *28*, 391–406. [CrossRef] [PubMed]

45. Hsieh, T.C.; Ma, K.H.; Chao, A. iNEXT: An R package for rarefaction and extrapolation of species diversity (Hill numbers). *Methods Ecol. Evol.* **2016**, *7*, 1451–1456. [CrossRef]

46. Chao, A.; Gotelli, N.J.; Hsieh, T.C.; Sander, E.L.; Ma, K.H.; Colwell, R.K.; Ellison, A.M. Rarefaction and extrapolation with Hill, numbers: A framework for sampling and estimation in species diversity studies. *Ecol. Monogr.* **2014**, *84*, 45–67. [CrossRef]

47. Chao, A.; Jost, L. Coverage-based rarefaction and extrapolation: Standardizing samples by completeness rather than size. *Ecology* **2012**, *93*, 2533–2547. [CrossRef]

48. Oksanen, J.; Blanchet, F.G.; Friendly, M.; Kindt, R.; Legendre, P.; McGlinn, D.; Minchin, P.R.; O'Hara, R.B.; Simpson, G.L.; Solymos, P.; et al. Vegan: Community Ecology Package. R Package Version 2.5-7. 2020. Available online: https://CRAN.R-project.org/package=vegan (accessed on 30 December 2021).

49. Clarke, K.R. Non-parametric multivariate analyses of changes in community structure. *Aust. J. Ecol.* **1993**, *18*, 117–143. [CrossRef]

50. Balogová, M.; Miková, E.; Orendáš, P.; Uhrin, M. Trophic spectrum of adult *Salamandra salamandra* in the Carpathians with the first note on food intake by the species during winter. *Herpetol. Notes* **2015**, *8*, 371–377.

51. Piñol, J.; Senar, M.A.; Symondson, W.O.C. The choice of universal primers and the characteristics of the species mixture determine when DNA metabarcoding can be quantitative. *Mol. Ecol.* **2019**, *28*, 407–419. [CrossRef]

52. Taberlet, P.; Bonin, A.; Zinger, L.; Coissac, E. DNA sequencing. In *Environmental DNA: For Biodiversity Research and Monitoring*; Oxford University Press: Oxford, UK, 2018. [CrossRef]

53. Andújar, C.; Creedy, T.J.; Arribas, P.; López, H.; Salces-Castellano, A.; Pérez-Delgado, A.J.; Vogler, A.P.; Emerson, B.C. Validated removal of nuclear pseudogenes and sequencing artefacts from mitochondrial metabarcode data. *Mol. Ecol. Resour.* **2021**, *21*, 1772–1787. [CrossRef]

54. Weisrock, D.W.; Hime, P.M.; Nunziata, S.O.; Jones, K.S.; Murphy, M.O.; Hotaling, S.; Kratovil, J.D. Surmounting the Large-Genome "Problem" for Genomic Data Generation in Salamanders. In *Population Genomics: Wildlife*; Hohenlohe, P.A., Rajora, O.P., Eds.; Springer International Publishing AG: Cham, Switzerland, 2018. [CrossRef]

55. Russell, J.C.; Kueffer, C. Island Biodiversity in the Anthropocene. *Annu. Rev. Environ. Resour.* **2019**, *44*, 31–60. [CrossRef]

56. Steinmetz, R.; Seuaturien, N.; Intanajitjuy, P.; Inrueang, P.; Prempree, K. The effects of prey depletion on dietary niches of sympatric apex predators in Southeast Asia. *Integr. Zool.* **2021**, *16*, 19–32. [CrossRef] [PubMed]

57. Ferenti, S.; David, A.; Nagy, D. Feeding-behaviour responses to anthropogenic factors on *Salamandra salamandra* (Amphibia, Caudata). *Biharean Biol.* **2010**, *4*, 139–143.

58. Van Gestel, C.; Loureiro, S.; Idar, P. Terrestrial isopods as model organisms in soil ecotoxicology: A review. *ZooKeys* **2018**, *801*, 127–162. [CrossRef] [PubMed]

59. Kuzmin, S.L. Feeding ecology of *Salamandra* and *Mertensiella*: A review of data and ontogenetic evolutionary trends. *Mertensiella* **1994**, *4*, 271–286.

60. Mulder, K.P.; Alarcón-Ríos, L.; Nicieza, A.G.; Fleischer, R.C.; Bell, R.C.; Velo-Antón, G. Independent evolutionary transitions to pueriparity across multiple timescales in the viviparous genus *Salamandra*. *Mol. Phylogenet. Evol.* **2022**, *167*, 107347. [CrossRef]

61. Alarcón-Ríos, L.; Nicieza, A.G.; Kaliontzopoulou, A.; Buckley, D.; Velo-Antón, G. Evolutionary history and not heterochronic modifications associated with viviparity drive head shape differentiation in a reproductive polymorphic species, *Salamandra salamandra*. *Evol. Biol.* **2020**, *47*, 43–55. [CrossRef]
62. Steinfartz, S.; Weitere, M.; Tautz, D. Tracing the first step to speciation: Ecological and genetic differentiation of a salamander population in a small forest. *Mol. Ecol.* **2007**, *16*, 4550–4561. [CrossRef]
63. Manenti, R.; Denoël, M.; Ficetola, G.F. Foraging plasticity favours adaptation to new habitats in fire salamanders. *Anim. Behav.* **2013**, *86*, 375–382. [CrossRef]
64. Velo-Antón, G.; Cordero-Rivera, A. Ethological and phenotypic divergence in insular fire salamanders: Diurnal activity mediated by predation? *Acta Ethol.* **2017**, *20*, 243–253. [CrossRef]

diversity

Article

Ecological Observations on Hybrid Populations of European Plethodontid Salamanders, Genus *Speleomantes*

Enrico Lunghi [1,2,3,4,*], Fabio Cianferoni [2,5], Stefano Merilli [6], Yahui Zhao [1,*], Raoul Manenti [7,8], Gentile Francesco Ficetola [7,9] and Claudia Corti [2]

1   Key Laboratory of the Zoological Systematics and Evolution, Institute of Zoology, Chinese Academy of Sciences, Beichen West Road 1, Beijing 100101, China
2   Museo di Storia Naturale dell'Università di Firenze, Museo "La Specola", Via Romana 17, 50125 Firenze, Italy; cianferoni.fabio@gmail.com (F.C.); claudia.corti@unifi.it (C.C.)
3   Natural Oasis, Via di Galceti 141, 59100 Prato, Italy
4   Unione Speleologica Calenzano, Piazza della Stazione 1, Calenzano, 50041 Firenze, Italy
5   Istituto di Ricerca sugli Ecosistemi Terrestri (IRET), Consiglio Nazionale delle Ricerche (CNR), Sesto Fiorentino, 50019 Firenze, Italy
6   Speleo Club Firenze via Guardavia 8, 50143 Firenze, Italy; s.merilli@virgilio.it
7   Department of Environmental Science and Policy, Universitòà degli Studi di Milano, Via Celoria, 26, 20133 Milan, Italy; raoulmanenti@gmail.com (R.M.); francesco.ficetola@gmail.com (G.F.F.)
8   Laboratorio di Biologia Sotterranea "Enrico Pezzoli", Parco Regionale del Monte Barro, 23851 Galbiate, Italy
9   Université Grenoble Alpes, Université Savoie Mont Blanc, CNRS, LECA, Laboratoire d'Ecologie Alpine, F-38000 Grenoble, France
*   Correspondence: enrico.arti@gmail.com (E.L.); zhaoyh@ioz.ac.cn (Y.Z.)

Citation: Lunghi, E.; Cianferoni, F.; Merilli, S.; Zhao, Y.; Manenti, R.; Ficetola, G.F.; Corti, C. Ecological Observations on Hybrid Populations of European Plethodontid Salamanders, Genus *Speleomantes*. *Diversity* **2021**, *13*, 285. https://doi.org/10.3390/d13070285

Academic Editors: Salvidio Sebastiano and Michael Wink

Received: 25 May 2021
Accepted: 17 June 2021
Published: 23 June 2021

Publisher's Note: MDPI stays neutral with regard to jurisdictional claims in published maps and institutional affiliations.

Abstract: *Speleomantes* are the only plethodontid salamanders present in Europe. Multiple studies have been performed to investigate the trophic niche of the eight *Speleomantes* species, but none of these studies included hybrid populations. For the first time, we studied the trophic niche of five *Speleomantes* hybrid populations. Each population was surveyed twice in 2020, and stomach flushing was performed on each captured salamander; stomach flushing is a harmless technique that allows stomach contents to be inspected. We also assessed the potential divergence in size and body condition between natural and introduced hybrids, and their parental species. Previously collected data on *Speleomantes* were included to increase the robustness of these analyses. In only 33 out of 134 sampled hybrid *Speleomantes* we recognized 81 items belonging to 11 prey categories. The frequency of empty stomachs was higher in females and individuals from natural hybrid populations, whereas the largest number of prey was consumed by males. We compared the total length and body condition of 685 adult salamanders belonging to three types of hybrids and three parental (sub)species. Three group of salamanders (one hybrid and two parental species) showed significantly larger size, whereas no difference in body condition was observed. This study provided novel ecological information on *Speleomantes* hybrid populations. We also provided insights into the potential divergence between hybrids and parental species in terms of size and body condition. We discuss our findings, and formulate several hypotheses that should be tested in the future.

Keywords: *Speleomantes*; *Hydromantes*; trophic niche; body condition; cave biology; biospeleology; parental species; diet; size; capture-mark-recapture

1. Introduction

European cave salamanders of the genus *Speleomantes* are the only plethodontids living in the Europe, and are almost all endemic to Italy [1]. Seven of the eight *Speleomantes* species (*S. ambrosii*, *S. italicus*, *S. flavus*, *S. supramontis*, *S. imperialis*, *S. sarrabusensis*, *S. genei*) live exclusively in Italy, while the range of one species (*S. strinatii*) extends to part of French Provence [1]. Each species is distributed in a well-defined area, and no range overlap exists; *Speleomantes* distribution is likely shaped by geomorphology [2,3]. For *S.*

ambrosii, the river Magra marks the separation of the two allopatric subspecies: the western *S. ambrosii ambrosii* and the eastern *S. a. bianchii* [1,4]. Although some phenotypic variability can be observed among *Speleomantes* [5–7], a valid method to phenotypically distinguish among species/subspecies is still lacking, thus their identification is based mostly on geography [1,3]. The southern distribution of the latter reaches the norther limit of *S. italicus*, creating a narrow contact zone in which viable hybrid populations are found [1,8] (see Figure 1 of Ref. [8]) These natural hybrid populations exhibit some genetic divergence, which is mainly influenced by the relative abundance of individuals of each parental species: in the northern part of the range there are populations of *S. a. bianchii* introgressed with *S. italicus*, whereas in the southern area, the opposite occurs [8]. However, hybrids do not show a clear divergence of phenotypic characters from their parental species; therefore, hybrids can also only be recognized on the basis of their geographic distribution [8,9]. In addition to this natural hybrid zone, hybrids between *S. ambrosii* and *S. italicus* are also related to a human-mediated translocation. In 1983, for scientific purposes individuals of both *S. italicus* and *S. a. ambrosii* were introduced into a natural cave in southern Tuscany, outside the natural range of *Speleomantes* [1,10]. Thirty years after its introduction, the population was genetically characterized and none of the individuals had a pure genotype. For 77% of individuals, the majority of alleles (>75%) matched alleles specific to *S. a. ambrosii*; for 6% of individuals, the majority of alleles matched those of *S. italicus*; and for 16% of individuals, alleles of *S. a. ambrosii* and of *S. italicus* were recombined [11]. However, the lack of ecological information on this population prevents us from evaluating the potential difference between these non-natural hybrids and the hybrids living in the natural hybrid zone, and to assess the trophic relationships between these populations and the local fauna [8,12]. Here we provide the first assessment of the size, body condition, and diet of both natural and introduced hybrid populations of *Speleomantes*.

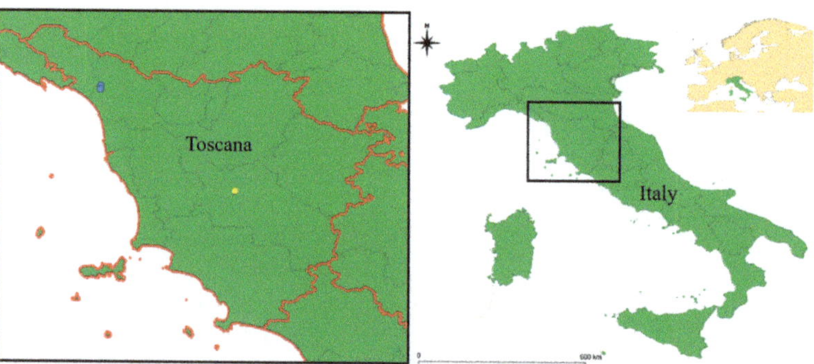

Figure 1. Map indicating the sampled populations. Blue labels indicate natural hybrid populations, whereas yellow indicates those introduced outside the natural distribution of *Speleomantes*. The coordinates of the surveyed sites are not provided to increase species protection [13].

2. Materials and Methods

In 2020 we surveyed five hybrid populations of *Speleomantes* that inhabit natural caves (Figure 1); each population was surveyed twice, once before (29 June–14 July) and after (3–10 September) the aestivation period [1].

Three populations are located within the natural hybrid zone between *S. italicus* and *S. ambrosii bianchii* occurring in north-western Tuscany (Province of Lucca); two populations included introgressed *S. italicus* with >10% of *S. ambrosii* alleles, whereas the other is a *S. ambrosii* population introgressed with >10% of *S. italicus* alleles [8]. The natural hybrid populations were selected following the genetic characterization of Ruggi, Cimmaruta, Forti and Nascetti [9]. The two other hybrid populations are found in southern Tuscany (Province of Siena), where cave salamanders are not native and have been introduced [1].

One of these two populations was discovered during this study. In the surroundings of the cave where *Speleomantes* were introduced [10], some authors (CC, EL, SM) have found and explored another cave about 53 m in a straight line from the previous one; because several *Speleomantes* were observed during the exploration, the individuals from this second cave were considered to be an additional population [14]. Surveys and animal handling were performed by taking measures to avoid the spread of pathogens (i.e., using disposable gloves and disinfecting boots and equipment after each survey). During each survey, all captured salamanders were placed in a perforated plastic box (60 × 40 × 20 cm). After salamanders were captured, we recorded the following data in sequence: individual sex (males were recognized by the presence of the distinctive mental gland, females were salamanders with SVL $\geq$ 40 mm without the mental gland, and remaining salamanders were considered juveniles) [1]; weight (using a digital scale, 0.01 g); a photo was taken from the dorsal view of individuals positioned along a reference card [6]; the harmless stomach flushing technique was used to evaluate *Speleomantes* foraging activity [15]. All salamanders were released at their collection points. The photos were analyzed with ImageJ software to measure the total length (TL) of the salamanders and estimate the snout-vent length (SVL) [16]. Stomach contents were preserved in a 75% ethanol solution and subsequently observed under an optical microscope, where prey items were recognized and counted following [17]. When we were unable to recognize any item at the order level, we considered the content as "unidentifiable"; when no remains were found, the stomach was considered "empty". The dorsal pattern of salamanders was used for individual recognition [18].

We used binomial Generalized Linear Mixed Models (GLMMs) to assess the potential effects that the considered variables may have on the stomach condition. Four individuals were captured twice, so we only used data from their first capture. We used stomach condition (empty/full) as the dependent variable, and the independent variables were salamander sex (male, female, juvenile) and hybrid identity (natural = *S. a. bianchii* × *S. italicus*; introduced = *S. a. ambrosii* × *S. italicus*); because the frequency of empty stomach changes over time [19], the survey month was added, together with the identity of the cave, as a random variable. Similarly, we used a GLMM to assess the potential correlation between the number of recognized prey and the independent variables mentioned above; cave and individual identity were still used as random variables.

We also performed a comprehensive analysis by combining these data with previously published datasets [5,8,20] to evaluate differences in size and body condition between hybrids and parental species. In these previous studies, the threshold to discriminate between adults and juvenile was set on the base of the smallest TL measured among males, which was 69 mm; in this study the smallest male had a SVL of 44 mm and TL of 69 mm. For each site, only the survey with the highest number of measured salamanders was considered, with the exception of hybrid data, in which pattern recognition [18] allowed the inclusion of individuals captured for the first time during the second survey. We used a Linear Mixed Model (LMM) to evaluate the potential differences in adult size (TL) between different groups of salamanders (*S. a. ambrosii*, *S. a. bianchii*, *S. italicus*, *S. a. ambrosii* × *S. italicus*, *S. a. bianchii* × *S. italicus*). A Shapiro–Wilk test showed a non-normal distribution of data related to *Speleomantes* size (log-transformed TL, $W = 0.98$, $p < 0.001$); however, LMM is appropriate for the analysis of non-normal distributed data [21]. The log-transformed TL was used as a dependent variable, and the group of salamanders as the independent variable. Considering the natural divergence in maximum size occurring between males and females [1], the sex of salamanders was used as a random factor. We used the Residual Index (RI) as a proxy of the body condition of the salamanders; this index provides information on the difference between the observed and the expected body mass [22,23]. To calculate the RI, we first log-transformed weight and TL, and then extracted residuals from the regression analysis for each species/hybrid group, in order to avoid bias due to different size [22,23]. *Speleomantes* body condition peaks during the foraging periods (i.e., when precipitation is higher and temperature relatively cold), and is poorest during inactivity periods (i.e., when climatic conditions are too hot and dry), when the salamanders mostly consume energy that was previously stored [1,24,25]. Therefore, in this analysis we only used data collected before the aestivation period (June–July). We used an LMM to assess the potential correlation between the RI (dependent variable) and two independent variables: the sex of the salamanders and the species group. Considering that data were collected over different periods, we included the month and year of the survey in addition to the identity of the cave, as random factors. The significance of GLMM and LMM variables was tested with a likelihood ratio test.

3. Results

In the hybrid populations, we obtained 138 salamander detections corresponding to 134 individuals; four individuals (one male and three females) were observed twice. The size of the recaptured individuals (both SVL and TL) did not change between the two surveys: the difference between the first and second measurement was <2 mm; this difference is comparable to measurement error [16]. Most of the individuals sampled (90) had an empty stomach. The frequency of empty stomachs was significantly different between sexes ($\chi^2 = 7.31$, df = 2, $p = 0.026$) and type of hybrid ($\beta = 4.41$, SE = 1.14, $\chi^2 = 14.91$, df = 1, $p < 0.001$); the frequency of empty stomachs was higher in females and individuals from natural hybrid populations. Eleven individuals had stomach contents in a state of advanced digestion and, therefore, the contents were considered not identifiable. We were able to recognize a total of 81 prey items from 33 individuals; the recognized prey belonged to 11 different categories: Sarcoptiformes (1), Mesostigmata (1), Araneae (4), Pseudoscorpiones (4), Polydesmida (3), Isopoda (8), Hemiptera (1), Hymenoptera (1), Coleoptera (3), Coleoptera_larva (1), and Diptera (55). Diptera were the most consumed prey: they were observed in 32 individuals, representing 67% of recognized prey. The number of prey consumed was significantly affected by the sex of salamanders ($\chi^2 = 10.74$, df = 2, $p = 0.005$); the largest number of prey was consumed by males.

When we combined the data of hybrid populations with those from previous surveys, we obtained data on the size and body condition of 678 salamanders (additional details in Figure 2). The maximum and average (±SD) size (TL) measured for the salamander groups considered were: *S. a. ambrosii*, females max. 125 and average 91 (±14) mm, and males max. 105 and average 92 (±6) mm; *S. a. bianchii*, females max. 114 and average 89 (±12) mm, and males max. 124 and average 105 (±12) mm; *S. italicus*, females max. 120 and average 94 (±13) mm, and males max. 118 and average 101 (±8) mm; *S. a. bianchii* × *S. italicus*, females max. 119 and average 90 (±15) mm, and males max. 119 and average 95 (±8) mm; *S. a. ambrosii* × *S. italicus*, females max. 134 and average 98 (±20) mm, and males max. 108 and average 103 (±2) mm. The size of the adult salamanders significantly differed between the groups ($F_{4,506} = 4.61$, $p = 0.001$); *S. italicus*, *S. a. bianchii*, and *S. a. ambrosii* × *S. italicus* hybrids had the largest size (Figure 2). No significant differences in body condition were observed between sexes ($F_{2,641} = 0.69$, $p = 0.5$) or species groups ($F_{4,25} = 0.16$, $p = 0.956$).

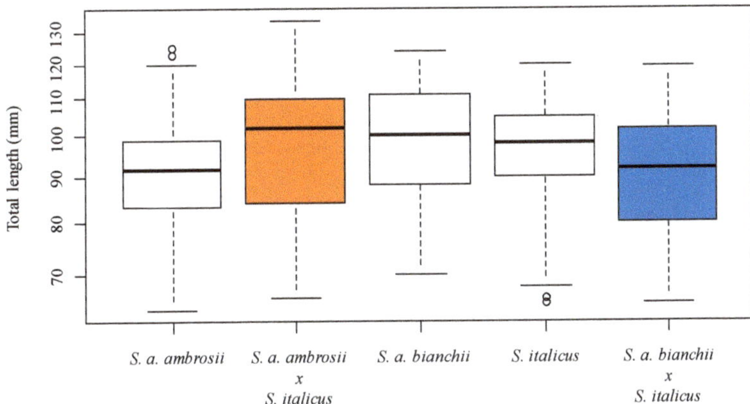

Figure 2. Boxplots showing the average size (TL) of adult *Speleomantes* considered in this study: *S. a. ambrosii* (163), *S. a. ambrosii* × *S. italicus* hybrids (28), *S. a. bianchii* (46), *S. italicus* (150), *S. italicus* × *S. a. bianchii* hybrids (125). The red box indicates introduced hybrids, blue natural hybrids, while in white the parental species.

4. Discussion

Studies on hybrid populations of *Speleomantes* have been extremely limited [8,11], leaving their morphological and ecological traits largely unknown. Despite our limited sample size, especially in regards to diet, we were able to obtain interesting information to provide the basis for future research. For example, we provide the first documentation relating to the natural colonization of a new environment performed by the introduced *Speleomantes*. Forty years after their release [10] the salamanders managed to occupy at least a nearby subterranean environment with a stable population. It is therefore of great importance to regularly monitor these hybrid populations to see if they will continue to extend their distribution.

The trophic niche of *Speleomantes* is subject to temporal variations [19,26] and is therefore strongly influenced by the fluctuating availability of potential prey [27]. With this study we were able to gather only limited information on the diet of *Speleomantes* hybrids. Most of the prey consumed were Diptera, which can be highly abundant near the entrance of subterranean environments, especially during the hot season [28,29]. This is in agreement with previous studies on the *Speleomantes* diet [16,30,31]. We also observed generally higher male foraging activity, whereas natural hybrids appear to forage less. Natural hybrids are distributed in a small area of the Apuan Alps (north-western Tuscany), a territory characterized by different lithology and vegetation compared to that of other studied salamanders [1,32,33]. Diverging environmental characteristics may offer different prey availability or simply alter the peak of foraging activity in these *Speleomantes* populations [24,27]. Therefore, further and prolonged investigations on multiple hybrid populations are needed to better delineate their trophic niche [17], evaluate potential divergences [27,30], and evaluate whether they increase foraging opportunities by expanding their microclimatic niche.

A further limitation of this study is the lack of genetic information on hybrid *Speleomantes* populations. Hybrid populations can undergo complex evolutionary phenomena, for example, with selective advantages for some components of their genome. To date, the available information on the genetic characteristics of hybrid populations is limited to allozyme data, which cannot capture the complexity of genomic processes [8]. Collecting genome-wide information on salamanders is challenging because their very large genome size makes the application of different genomic tools, such as RAD sequencing, problematic. However, recent developments can allow more affordable analysis of genome-wide variation, even in species with large genomes, and in the coming years this could improve our understanding of evolutionary processes affecting these populations [31,32].

In this study we observed larger size in introduced hybrids (*S. a. ambrosii* × *S. italicus*), *S. italicus*, and the *S. a. bianchii* subspecies. We do not have data on the potential genetic effect on the size of hybrid populations. However, ecological conditions are more likely to play an important role in driving the evolution of this adaptive trait. It has been shown that *Speleomantes* are in thermal equilibrium with their surrounding environment, and that larger individuals require more time to thermoconform, and are thus potentially able to exploit a less suitable microclimate for a longer period [8,20,33], allowing them to extend their activity period and increase foraging [24]. The larger size observed in introduced hybrids may be the result of a warmer and drier local climate, whereas the smaller size of natural hybrids may have been determined by the local colder and moister climatic conditions. This remains a hypothesis that deserves further investigation. Furthermore, introduced hybrids showed a lower frequency of empty stomachs compared to other population; thus, it is possible that higher food availability allows for better feeding and growth. However, this hypothesis is not supported by the lack of differences in the body condition index. Considering the larger size of the introduced hybrids (and the associated better tolerance of harsher conditions), we recommend regular monitoring of these populations to better control their range expansion, especially in the context of global warming [34,35].

Author Contributions: E.L. conceived the study, analysed the data, and prepared the figures; E.L., S.M., R.M., G.F.F. and C.C. performed field surveys; E.L. and F.C. performed the recognition of stomach contents; E.L., F.C., S.M., Y.Z., R.M., G.F.F. and C.C. reviewed and edited the manuscript. All authors have read and agreed to the published version of the manuscript.

Funding: The APC was funded by National Natural Science Foundation of China (NSFC-31972868).

Institutional Review Board Statement: This study was authorized by the Italian Ministry of the Environment (Ministero dell'Ambiente e della Tutela del Territorio e del Mare, PNM prot. n.56097/T-A31 of 13/11/2017 and integration, prot. n. 7513 of 17/02/2020).

Informed Consent Statement: Not applicable.

Data Availability Statement: Data analysed in this study can be found at this link https://figshare.com/s/387b39c7de65dfc32298.

Acknowledgments: This study was conducted under the auspices of the Chinese Cavefish Working Group. Enrico Lunghi is supported by the Chinese Academy of Sciences President's International Fellowship Initiative for postdoctoral researchers. Caves' coordinate may be provided upon justified requests.

Conflicts of Interest: The authors declare no conflict of interest.

References

1. Lanza, B.; Pastorelli, C.; Laghi, P.; Cimmaruta, R. A review of systematics, taxonomy, genetics, biogeography and natural history of the genus *Speleomantes* Dubois, 1984 (Amphibia Caudata Plethodontidae). *Atti. Mus Civ. Stor. Nat. Trieste* **2006**, *52*, 5–135.
2. Lanza, B.; Andreone, F.; Bologna, M.A.; Corti, C.; Razzetti, E. *Fauna d'Italia. Amphibia*; Calderini: Bologna, Italy, 2006.
3. Chiari, Y.; van der Meijden, A.; Mucedda, M.; Lourenço, J.M.; Hochkirch, A.; Veith, M. Phylogeography of Sardinian cave salamanders (genus *Hydromantes*) is mainly determined by geomorphology. *PLoS ONE* **2012**, *7*, e32332. [CrossRef]
4. Carranza, S.; Romano, A.; Arnold, E.N.; Sotgiu, G. Biogeography and evolution of European cave salamanders, *Hydromantes* (Urodela: Plethodontidae), inferred from mtDNA sequences. *J. Biogeogr.* **2008**, *35*, 724–738. [CrossRef]
5. Lunghi, E.; Cianferoni, F.; Giachello, S.; Zhao, Y.; Manenti, R.; Corti, C.; Ficetola, G.F. Updating salamander datasets with phenotypic and stomach content information for two mainland *Speleomantes*. *Sci. Data* **2021**, *8*, 150. [CrossRef]
6. Lunghi, E.; Giachello, S.; Zhao, Y.; Corti, C.; Ficetola, G.F.; Manenti, R. Photographic database of the European cave salamanders, genus *Hydromantes*. *Sci. Data* **2020**, *7*, 171. [CrossRef] [PubMed]
7. Lanza, B.; Caputo, V.; Nascetti, G.; Bullini, L. Morphologic and genetic studies of the European plethodontid salamanders: Taxonomic inferences (genus *Hydromantes*). *Monogr. Mus. Reg. Sci. Nat. Torino* **1995**, *16*, 1–366.
8. Ficetola, G.F.; Lunghi, E.; Cimmaruta, R.; Manenti, R. Transgressive niche across a salamander hybrid zone revealed by microhabitat analyses. *J. Biogeogr.* **2019**, *46*, 1342–1354. [CrossRef]

9.   Ruggi, A.; Cimmaruta, R.; Forti, G.; Nascetti, G. Preliminary study of a hybrid zone between *Speleomantes italicus* Dunn 1923 and *S. ambrosii* Lanza 1955 on the Apuan Alps, using RLFP analysis. *Ann. Mus. Civ. Stor. Nat. Giacomo Doria Genova* **2005**, *97*, 135–144.

10. Forti, G.; Lanza, B.; Cimmaruta, R.; Nascetti, G. An experiment of artificial syntopy ex situ between *Speleomantes italicus* (Dunn, 1923) and *S. ambrosii ambrosii* (Lanza, 1955) (Amphibia, Plethodontidae). *Ann. Mus. Civ. Stor. Nat. Giacomo Doria Genova* **2005**, *97*, 123–133.

11. Cimmaruta, R.; Forti, G.; Lucente, D.; Nascetti, G. Thirty years of artificial syntopy between *Hydromantes italicus* and *H. ambrosii ambrosii* (Amphibia, Plethodontidae). *Amphib.-Reptil.* **2013**, *34*, 413–420. [CrossRef]

12. Lunghi, E.; Guillaume, O.; Blaimont, P.; Manenti, R. The first ecological study on the oldest allochthonous population of European cave salamanders (*Hydromantes* sp.). *Amphib.-Reptil.* **2018**, *39*, 113–119. [CrossRef]

13. Lunghi, E.; Corti, C.; Manenti, R.; Ficetola, G.F. Consider species specialism when publishing datasets. *Nat. Ecol. Evol.* **2019**, *3*, 319. [CrossRef] [PubMed]

14. Lunghi, E.; Bruni, G. Long-term reliability of visual implant elastomers in the Italian cave salamander (*Hydromantes italicus*). *Salamandra* **2018**, *54*, 283–286.

15. Crovetto, F.; Romano, A.; Salvidio, S. Comparison of two non-lethal methods for dietary studies in terrestrial salamanders. *Wildl. Res.* **2012**, *39*, 266–270. [CrossRef]

16. Lunghi, E.; Giachello, S.; Manenti, R.; Zhao, Y.; Corti, C.; Ficetola, G.F.; Bradley, J.G. The post hoc measurement as a safe and reliable method to age and size plethodontid salamanders. *Ecol. Evol.* **2020**, *10*, 11111–11116. [CrossRef] [PubMed]

17. Lunghi, E.; Cianferoni, F.; Ceccolini, F.; Mulargia, M.; Cogoni, R.; Barzaghi, B.; Cornago, L.; Avitabile, D.; Veith, M.; Manenti, R.; et al. Field-recorded data on the diet of six species of European *Hydromantes* cave salamanders. *Sci. Data* **2018**, *5*, 180083. [CrossRef]

18. Lunghi, E.; Romeo, D.; Mulargia, M.; Cogoni, R.; Manenti, R.; Corti, C.; Ficetola, G.F.; Veith, M. On the stability of the dorsal pattern of European cave salamanders (genus *Hydromantes*). *Herpetozoa* **2019**, *32*, 249–253. [CrossRef]

19. Lunghi, E.; Cianferoni, F.; Ceccolini, F.; Veith, M.; Manenti, R.; Mancinelli, G.; Corti, C.; Ficetola, G.F. What shapes the trophic niche of European plethodontid salamanders? *PLoS ONE* **2018**, *13*, e0205672. [CrossRef] [PubMed]

20. Ficetola, G.F.; Lunghi, E.; Canedoli, C.; Padoa-Schioppa, E.; Pennati, R.; Manenti, R. Differences between microhabitat and broad-scale patterns of niche evolution in terrestrial salamanders. *Sci. Rep.* **2018**, *8*, 10575. [CrossRef]

21. Schielzeth, H.; Dingemanse, N.J.; Nakagawa, S.; Westneat, D.F.; Allegue, H.; Teplitsky, C.; Réale, D.; Dochtermann, N.A.; Garamszegi, L.Z.; Araya-Ajoy, Y.G. Robustness of linear mixed-effects models to violations of distributional assumptions. *Methods Ecol. Evol.* **2020**, *11*, 1141–1152. [CrossRef]

22. Labocha, M.K.; Schutz, H.; Hayes, J.P. Which body condition index is best? *Oikos* **2014**, *123*, 111–119. [CrossRef]

23. Băncilă, R.I.; Hartel, T.; Băncilă, R.I.; Smets, J.; Cogălniceanu, D. Comparing three body condition indices in amphibians: A case study of yellow-bellied toad *Bombina variegata*. *Amphib.-Reptil.* **2010**, *31*, 558–562. [CrossRef]

24. Lunghi, E.; Manenti, R.; Mulargia, M.; Veith, M.; Corti, C.; Ficetola, G.F. Environmental suitability models predict population density, performance and body condition for microendemic salamanders. *Sci. Rep.* **2018**, *8*, 7527. [CrossRef] [PubMed]

25. Lunghi, E.; Manenti, R.; Ficetola, G.F. Seasonal variation in microhabitat of salamanders: Environmental variation or shift of habitat selection? *PeerJ* **2015**, *3*, e1122. [CrossRef]

26. Salvidio, S.; Romano, A.; Oneto, F.; Ottonello, D.; Michelon, R. Different season, different strategies: Feeding ecology of two syntopic forest-dwelling salamanders. *Acta Oecol.* **2012**, *43*, 42–50.

27. Lunghi, E.; Manenti, R.; Cianferoni, F.; Ceccolini, F.; Veith, M.; Corti, C.; Ficetola, G.F.; Mancinelli, G. Interspecific and inter-population variation in individual diet specialization: Do environmental factors have a role? *Ecology* **2020**, *101*, e03088. [CrossRef]

28. Manenti, R.; Lunghi, E.; Ficetola, G.F. Distribution of spiders in cave twilight zone depends on microclimatic features and trophic supply. *Invertebr. Biol.* **2015**, *134*, 242–251. [CrossRef]

29. Lunghi, E.; Ficetola, G.F.; Zhao, Y.; Manenti, R. Are the neglected Tipuloidea crane flies (Diptera) an important component for subterranean environments? *Diversity* **2020**, *12*, 333. [CrossRef]

30. Lunghi, E.; Cianferoni, F.; Ceccolini, F.; Zhao, Y.; Manenti, R.; Corti, C.; Ficetola, G.F.; Mancinelli, G. Same diet, different strategies: Variability of individual feeding habits across three populations of Ambrosi's cave salamander (*Hydromantes ambrosii*). *Diversity* **2020**, *12*, 180. [CrossRef]

31. Pierson, T.W.; Fitzpatrick, B.M.; Camp, C.D. Genetic data reveal fine-scale ecological segregation between larval plethodontid salamanders in replicate contact zones. *Evol. Ecol.* **2021**, *35*, 309–322. [CrossRef]

32. Gargiulo, R.; Kull, T.; Fay, M.F. Effective double-digest RAD sequencing and genotyping despite large genome size. *Mol. Ecol. Resour.* **2021**, *21*, 1037–1055. [CrossRef] [PubMed]

33. Lunghi, E.; Manenti, R.; Canciani, G.; Scarì, G.; Pennati, R.; Ficetola, G.F. Thermal equilibrium and temperature differences among body regions in European plethodontid salamanders. *J. Therm. Biol.* **2016**, *60*, 79–85. [CrossRef] [PubMed]

34. Markle, T.M.; Kozak, K.H. Low acclimation capacity of narrow-ranging thermal specialists exposes susceptibility to global climate change. *Ecol. Evol.* **2018**, *8*, 4644–4656. [CrossRef]

35. Vieites, D.R.; Min, M.-S.; Wake, D.B. Rapid diversification and dispersal during periods of global warming by plethodontid salamanders. *Proc. Natl. Acad. Sci. USA* **2007**, *104*, 19903–19907. [CrossRef] [PubMed]

Article

A Midsummer Night's Diet: Snapshot on Trophic Strategy of the Alpine Salamander, *Salamandra atra*

Luca Roner [1], Andrea Costa [2], Paolo Pedrini [1], Giorgio Matteucci [3,4], Stefano Leonardi [5] and Antonio Romano [1,3,*]

1   MUSE—Museo delle Scienze, Sezione di Zoologia dei Vertebrati, Corso del Lavoro e della Scienza 3, 38122 Trento, Italy; lucaroner@gmail.com (L.R.); paolo.pedrini@muse.it (P.P.)
2   Dipartimento di Scienze della Terra, Ambiente e Vita, Università degli Studi di Genova, Corso Europa 26, I-16132 Genova, Italy; andrea-costa-@hotmail.it
3   Consiglio Nazionale delle Ricerche, Istituto per i Sistemi Agricoli e Forestali del Mediterraneo, Via Patacca, 84, 80056 Ercolano (NA), Italy; giorgio.matteucci@isafom.cnr.it
4   Consiglio Nazionale delle Ricerche, Istituto per la BioEconomia, Via Madonna del Piano, 10, 50019 Sesto Fiorentino (FI), Italy
5   Dipartimento di Scienze Chimiche, della Vita e della Sostenibilità Ambientale, Università degli studi di Parma, Parco Area delle Scienze, 11/a, 43124 Parma, Italy; stefano.leonardi@unipr.it
*   Correspondence: antonioromano71@gmail.com

Received: 30 March 2020; Accepted: 15 May 2020; Published: 17 May 2020

Abstract: Information on the trophic ecology of the Alpine salamander, *Salamandra atra*, is scattered and anecdotal. We studied for the first time the trophic niche and prey availability of a population from an area located in Italian Dolomites during the first half of August. Considering that *S. atra* is a typical nocturnal species, we collected food availability separately for diurnal and nocturnal hours. Our aims were: (i) to obtain information on the realized trophic niche; (ii) to provide a direct comparison between trophic strategy considering only nocturnal preys or considering all preys; (iii) to study trophic strategy of this species at the individual level. In two samplings nights we obtained prey from 50 individuals using stomach flushing technique. Trophic strategy was determined using the graphical Costello method and selectivity using the relativized electivity index. During the short timeframe of our sample, this salamander showed a generalized trophic strategy. The total trophic availability differed significantly from nocturnal availability. Interindividual diet variation is discussed in the light of the optimal diet theory. Finally, we highlighted that considering or not the activity time of the studied taxon and its preys may lead to a conflicting interpretation of the trophic strategies.

Keywords: amphibians; feeding ecology; individual specialization; resource selection; salamanders

---

1. Introduction

The ecological role of salamanders is often overlooked, despite the fact they can act as top predators in certain trophic webs [1]. They may represent a significant part of vertebrate biomass in North American ecosystems [2], and can reach high densities in North American [3] and European forests as well [4,5]. Salamanders mainly prey on invertebrates and play a key role in nutrient cycling [6]. Furthermore, those species characterized by a biphasic life cycle are also an important energy exchange vector among different habitat types [1]. For these reasons, dietary studies on salamanders are an indispensable tool for assessing their ecological role, and for planning future conservation measures [7]. At population level, the effect of the diet, together with other ecological factors and variables such as climate, predation, human pressure, stress, and disease (e.g., [8–11]), is considered of paramount importance to determine animals' abundance [12]. Considering the use of the trophic resources, populations may be assessed as generalist if they are composed by individuals feeding to the environmental availability of prey proportionally or as specialist if individuals select only a limited array of the available resource categories [13]. Amphibians in general, and salamanders in particular, are often seen as generalist, opportunistic predators that feed on a large range of prey [14]. However, some studies highlighted how salamanders actively select prey and may show diet specialization, at least under particular environmental conditions [15,16]. Moreover, while the realized trophic niche may result as the outcome of a generalist feeding strategy in a given population, it could be actually composed of specialist individuals consuming different resources [13,17]. Patterns of individual specialization, or interindividual diet variation, occur in hundreds of cases and in many taxa [18]. Interindividual diet variation indeed was observed in salamanders too, both at the postmetamorphic [16,19–21] and at the larval stage [22]. Individual specialization in salamanders is usually inferred by cross-sectional data, but also with longitudinal studies [20].

The Alpine salamander, *Salamandra atra* Laurenti 1768, is a widespread terrestrial salamander occurring in the Central and Eastern Alps, and in the Dinaric Alps where some isolated populations may be found [23,24]. Salamanders are abundant vertebrates in some environments and contribute to ecosystem resilience–resistance in several ways. One of these concerns their role as predators. The goal of this paper is to elucidate some aspects of the *Salamandra atra* as a predator in an alpine environment. There are few and mainly observational studies [25–28] on the feeding habits of the focal species, which were reviewed by Kuzmin [29]. Therefore, information on its trophic strategy is scattered and anecdotal [23,25–29]. So far, no fully quantitative information on the Alpine salamander diet comes from a single study [30]; that, however, was performed without taking into account trophic availability and feeding strategies. In the present paper, although during a short time frame about in the midseason of activity, we focused on the trophic strategy of this species at three different levels. At the top level, we defined the realized trophic niche by analyzing the mutual proportion of the preyed taxa. At the second level, we studied the trophic strategy in terms of prey selection, considering relationships between preyed taxa and their environmental availability, also taking into account the mainly nocturnal behavior of the Alpine salamander. Finally, we studied the trophic strategy of this species at the individual level.

2. Materials and Methods

2.1. Study Species

The Alpine salamander, *Salamandra atra*, is a fully terrestrial and viviparous salamander [24]. Mixed coniferous and deciduous forests, alpine meadows, and rocky tundra-like areas, mainly on limestone substrates, are the typical habitats. Activity pattern is concentrated in the warmest months (April–October) while in the rest of the year the salamanders are inactive at the ground surface [28].

2.2. Study Area

The study area (about 4000 m^2) is located in the Paneveggio–Pale di San Martino Natural Park (Northern Italy), at about 1850 m a.s.l., near the locality Malga Venegiotta (municipality of Tonadico; 46°18′48″ N, 11°48′53″ E). At macroscale, the area is characterized by open habitats (pastures, other grasslands, and rocky areas) mixed with coniferous woodland, which is dominated by European larch (*Larix decidua*) and Norway spruce (*Picea abies*). Following the fourth level Corine Land Cover nomenclature, the study area is classified as "Coniferous forests with discontinuous canopy on mire" (habitat 3.1.2.4). At smaller scale, the study area, where the sampling of salamanders and invertebrates was performed, is homogeneous and characterized by some coniferous trees mixed with small open habitat, such as rocks, grass, and dwarf shrub cushions dominated by *Erica carnea*. This site is especially suitable for sampling predators since at this place the detection probability of salamanders is high, as shown by a previous study [5].

2.3. Prey Availability (Potential Trophic Niche)

Ground-dwelling invertebrates were sampled by 20 pitfall traps. Each pitfall trap (500 cm^3) was partially filled with a killing/preserving solution (salty water with and addition of 500 mg of benzoic/acetic acid) [31,32]. Pitfall traps were active for four days immediately after salamander sampling. They were divided in two typologies: 10 diurnal traps (DT, which were active from 7:00 a.m. to 7:00 p.m.) and 10 nocturnal traps (NT; which were active from 7:00 p.m. to 7:00 a.m.). DT and NT were inactivated by covering them with a plastic lid when they did not have to capture invertebrates. Pitfall traps are widely used to measure diversity and abundance of ground-dwelling invertebrates, e.g., [33]. These traps, although they may overestimate mobile fauna [33,34], it is reasonable to assume that such bias would not be misleading in the assessment of prey availability since prey mobility increases the detection probability by amphibians [35,36]. Traps were placed in 10 different trapping points within the salamander sampling area separated by a minimum distance of 30 m. In each trapping point, a pair of two pitfall traps (one DT and one NT) was placed at 20–50 cm from each other. Thus, in each trapping point we were able to sample diurnal and nocturnal invertebrate separately. To prevent the accidental fall of the salamanders in the traps, a 20 mm-mesh rigid plastic net was placed at the entrance of the traps when they were activated. Invertebrates obtained from environmental sampling and from stomach contents were sorted, identified, and counted using a dissecting microscope and taxonomic keys. Since invertebrates obtained with stomach flushing are partly digested, all invertebrates, both from stomach contents and from environmental sampling, were generally determined at the Order level or higher, annotating the life stage (i.e., we distinguished larvae from adults) and radical differences in locomotion type (e.g., flying Hymenoptera were distinguished from walking ones, e.g., ants).

2.4. Sampling Predators

Sampling of salamanders occurred within an area of about 4000 m^2 in the first half of August 2018. Salamanders were sampled following rain and while active on the floor during two consecutive nights. They were transported to the laboratory, 5.5 km from the sampling site. Stomach contents were obtained by stomach flushing [37,38] performed by a single person using a 5 mL syringe [one injection per salamander] and a flexible soft plastic tube and preserved in 70% ethanol. Since there is a significant increase in digestion rate with increasing temperature [39], salamanders were stored at 5 °C in a refrigerator and they were flushed within three hours from capture [40,41]. A removal approach was used to avoid recaptures of the same individuals. Salamanders were photographed with a digital camera situated perpendicular to the dorsal surfaces of the animals. Digital photographs of salamanders were imported into the ImageJ® software program to measure their total length (TOTL, distance from the tip of the snout to the end of the tail). Sexes were distinguished by analysis of external secondary sexual characters; adult males have a prominent, swollen cloaca and are more slender than females. According to Klewen [27], we considered as "juveniles", which were excluded in the present study, those individuals without evident external secondary sexual characters and a TOTL smaller than 90 mm. All salamanders were returned to their original site within a maximum of 30 h from their capture.

2.5. Data Analysis

2.5.1. Realized Trophic Niche

The sex differentiation in diet was analyzed by means of analysis of similarity [ANOSIM], based on Bray–Curtis distance [42]. The diversities of prey taxa in salamander stomachs and in the environment, as well as the diversity of prey taxa in salamander stomachs of our population and the two populations studied by Fachbach et al. [30], were estimated through Simpson's index [1-D] and 95% confidence limits calculated by bootstrapping [43]. In fact, although our method (stomach flushing) and that used by Fachbach et al. ([30], stomach dissection) are different, these two methods provide comparable results [44]. Analyses were performed in the statistical package PAST [45]. Considering that prey availability is generally calculated on invertebrates captured within 24 hours for a few days, we compared the results of diurnal and nocturnal traps and the results of nocturnal traps versus the pooled results (i.e., diurnal plus nocturnal preys) by means of the diversity permutation test [9999 permutations].

2.5.2. Trophic Strategy

The use of prey types in relation to their abundance in the environment was estimated by means of the Vanderploeg and Scavia [46] relativized electivity index (E^*), which is strongly supported by comparative evaluations [47]:

$$E^* = (W_i - 1/n)/(W_i + 1/n)$$

where $W_i = (r_i/p_i)/(\Sigma\ r_i/p_i) - 1$, r_i is the relative abundance of prey i in the diet, p_i is the relative abundance of prey i in the environment, and n is the number of prey types. This index ranges from +1 (positive selection) to −1 (avoidance), while $E^* = 0$ indicated that prey items were consumed according to their availability. Since the index is particularly sensitive to the categories of prey with reduced environmental availability, and considering the low number of individuals within each prey taxon in our samples, the threshold electivity value (u) was considered only for prey type with more than three trapped individuals, calculating the fifth percentile of the absolute values of E^* [15,48]. The trophic strategy of the Alpine salamander was also analyzed with a modification of Costello's graphical representation [49,50]. According to this method, each prey type is plotted on a graph in which the x-axis is the prey frequency of occurrence (FO) in the predators' stomachs, and the y-axis is the prey-specific abundance (P_i), defined as the proportion of prey items (i), considering only all the prey items found in the individuals that consumed that specific prey type [50]. This graphical approach gives insights on the population feeding strategy: specialized (when some prey taxa have high P_i values and are projected in the upper part of the plot) vs. generalist (when all prey taxa have low P_i values and are projected in the lower part of the plot).

2.5.3. Interindividual Diet Variation

Interindividual diet variation for the study population was assessed by means of network analysis [18,51]. Within this approach, the interactions between individuals and resources are interpreted as a bipartite network where two sets of nodes, one representing individual salamanders and one representing prey types, are connected by links reflecting the utilization of each prey type by individuals [52–54]. Individual specialization with bipartite network is often investigated with qualitative data [52,54] that only represent the use of a resource. The use of weighted networks where the frequency of use of each resource is retained, however, may give better estimates of some network metrics; therefore, we decided to use this approach [53,55]. Within our network approach, we employed the degree of diet variation E, as proposed by Araujo et al. [51] to quantify the presence of interindividual diet variation. This index is based on the pairwise diet overlap between individuals and increases from 0 to 1 in presence of individual specialization [18,51]. Two other network metrics were calculated: nestedness and modularity. Nestedness is observed when individuals with the narrowest trophic niche consume a subset of the prey types used by the more generalist individuals. The latter is recorded when, within a population, it is possible to segregate some individuals in groups (modules) that share the same resources. We used a metric of nestedness based on overlap and decreasing fill (NODF; [56]), which ranges from 0 to 100 (minimum and maximum nestedness, respectively). Modularity Q was measured and modules within the population were identified (function computeModules in the R package Bipartite) using the Beckett's algorithm [57], which ranges from 0 to 1 (minimum and maximum modularity, respectively). Since some level of both specialization degree, nestedness and modularity may arise from stochastic processes and sampling bias, significance of these metrics was tested by comparing the observed value against the simulated distribution obtained from a null model with 9999 (999 for modularity) resamplings.

3. Results

3.1. Prey Availability (Potential Trophic Niche)

During this study, 19 taxa of invertebrates were captured for a total of 650 individuals, of which 395 and 255 individuals were captured in the diurnal and nocturnal traps, respectively (Table 1).

Table 1. Environmental availability of invertebrates and number of invertebrates preyed by *Salamandra atra* in the study site.

Invertebrate Taxa	Preyed Invertebrates	Environmental Availability of Invertebrates		
	Stomach Contents	*Diurnal Traps (DT)*	*Nocturnal Traps (NT)*	*Pool captures (DT + NT)*
Arachnida	12	37	17	54
Chilopoda	17	5	3	8
Coleoptera	13	23	6	29
Coleoptera larvae	8	6	7	13
Collembola	3	74	101	175
Dermaptera	0	1	1	2
Diptera	3	66	12	78
Diptera larvae	38	0	1	1
Formicidae	0	93	61	154
Hemiptera	0	13	6	19
Isopoda	9	6	14	20
Lepidoptera	0	2	1	3
Lepidoptera larvae	6	1	1	2
Mecoptera larvae	0	0	5	5
Mollusca	19	2	17	19
Oligochaeta	6	1	1	2
Orthoptera	0	2	0	2
Rynchota	0	2	0	2
winged Hymenoptera	2	4	1	5

Diversity index (Table 2) did not differ significantly between diurnal and total prey availability (i.e., DT + NT) (diversity permutation test, $p = 0.91$) while the difference between nocturnal and total prey availability was highly significant (diversity permutation test, $p < 0.01$). The analysis of similarities (ANOSIM) between total and nocturnal prey availability also showed significant differences (R = 0.29; $p < 0.01$).

Table 2. Diversity index of the available prey.

Simpson Diversity Index	1-D (95% C.I.)	Taxa
Diurnal traps (DT)	0.82 (0.80–0.83)	17
Nocturnal traps (NT)	0.76 (0.72–0.80)	17
Pooled captures (i.e., DT + NT)	0.82 (0.80–0.83)	19

3.2. Sampling Predators

Fifty adult salamanders (i.e., with total length longer than 90 mm) were captured (26 females, 24 males) and stomach flushed with 41 positive, 4 individuals without prey in stomach, and 5 individuals with only indeterminate items (portions and fragments of prey unrecognizable, which were not attributable to an exact number of prey). By sorting of stomach contents, 176 invertebrates were obtained with a total of 139 analyzable items (n = 26 indeterminate, n = 11 parasite nematode) with an average of 2.78 ± 5.05 preys/stomach (mean ± s.d.; n = 45; range 0–34. Parasite nematodes were excluded) (Table 1).

3.3. Realized Trophic Niche

There was no overall difference in the diet composition between the sexes (ANOSIM, n = 41; global $R = -0.049$, $p = 0.882$). The analysis of the trophic niche, using the modification of Costello's graphical method [49,50], showed that *Salamandra atra* exhibited a generalized trophic strategy (Figure 1). Almost all prey categories are located in the left lower part of the graph, with both FO and $P_i < 0.50$, with only Diptera (fly) larvae located in the upper left quadrant (Figure 1). Fly larvae are the most abundant prey category in the diet, but they are eaten by a low number of individuals (FO = 0.097), thus suggesting a generalized trophic strategy at the population level.

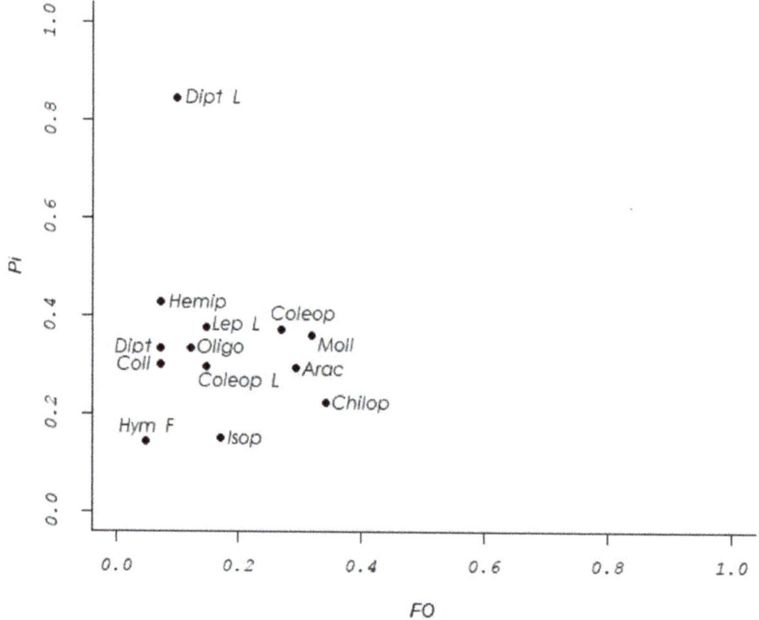

Figure 1. Modified Costello's plot [49,50], describing the trophic strategy of *Salamandra atra*. Legend: Arac = Aracnidae, Chilop = Chilopoda, Coleop = Coleoptera adults, Coleop_L = Coleoptera larvae, Coll = Collembola, Dipt = Diptera adults, Dipt_L = Diptera larvae, Hemip = Hemiptera, Hym_F = flying Hymenoptera, Isop = Isopoda, Lep_L = Lepidoptera larvae, Moll = Mollusca, Oligo = Oligochaeta.

The comparison of prey percentage in stomach of Alpine salamanders from our population and those of Fachbach et al. [30] is reported in Figure 2. Although some prey taxa were exclusive of a given population (e.g., Balttoidea, Dermaptera, Rhynchota), in the whole these three populations showed very similar diversity indices of the prey taxa (Simpson's index, 1-D, with 95% confidence limits, CL. Population 1: 1-D = 0.86, CL = 0.86–0.90; Population 2: 1-D = 0.85, CL = 0.83–0.90; Population 3: 1-D = 0.86, CL = 0.83–0.90).

3.4. Analysis of Trophic Strategy

Electivity index E^* was calculated using both nocturnal (Figure 3) and total (Figure 4) trophic availability. In both cases, E^* was negative for Hemiptera, Isopoda, Diptera, Aracnida, and Collembola, positive for Chilopoda, and proportional to environmental availability for Coleoptera larvae. Conversely, Mollusca resulted positively selected considering the total trophic availability, but they were randomly selected in the nocturnal availability. Worth noting is that E^* provided completely opposite results for Coleoptera, which were negatively selected considering total availability but positively considering only nocturnal traps. Therefore, the two trophic availabilities (nocturnal and diurnal), when analyzed separately provided different or even an opposite pattern of prey's selectivity. These different results arised also considering the two main prey types, as shown by the realized trophic niche. Finally, salamanders operated a strong negative selection on the more representative categories in the trophic availability, which are Collembola, Diptera, and Formicidae.

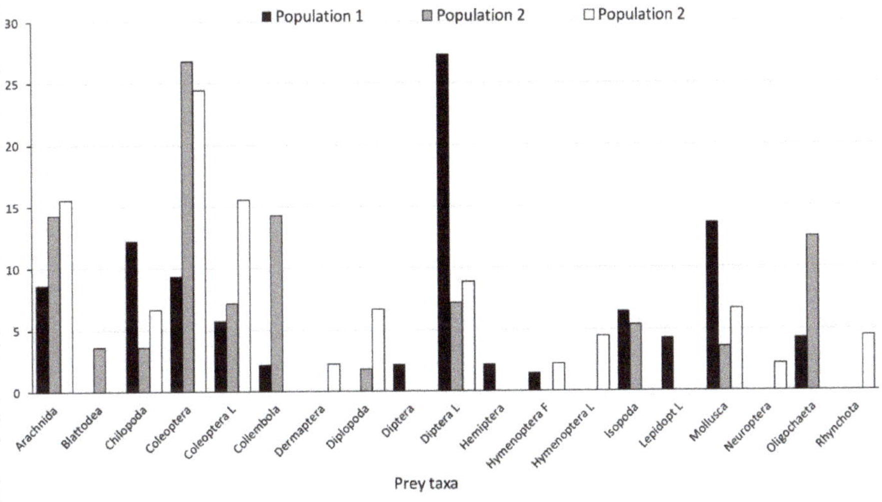

Figure 2. Comparison of prey percentage in stomach of Alpine salamanders from our population (Population 1, *n* = 41) and those of Fachbach et al. [30] (Population 2, from Gleinalmspeik, Germany, *n* = 15; Population 3, from Grimming, Germany, *n* = 26) [30]. The suffix L indicates the larval stage of the taxa; the suffix F (flying) indicates the winged taxa or the winged stage of the taxa.

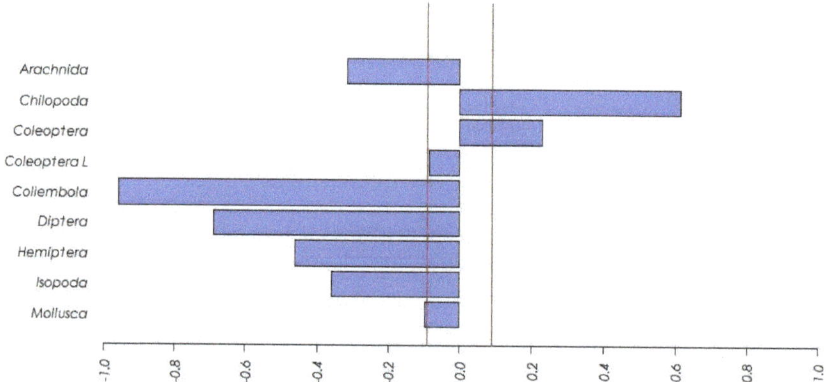

Figure 3. Relativized electivity index E* based on nocturnal trophic availability. Within the red vertical lines values are not statistically significant. The suffix L indicates the larval stage of the taxa.

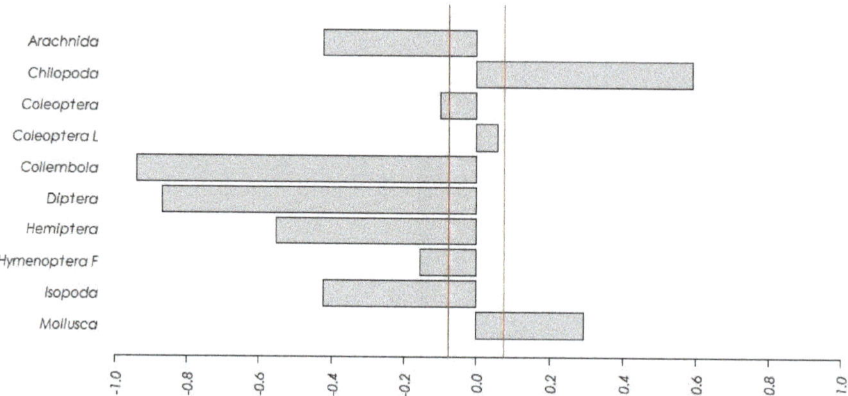

Figure 4. Relativized Electivity index E* based on total trophic availability. Within the red vertical lines values are not statistically significant. The suffix L indicates the larval stage of the taxa; the suffix F (flying) indicates the winged taxa or the winged stage of the taxa.

3.5. Analysis of Interindividual Diet Variation

The weighted bipartite network of individuals and resources is presented in Figure 5. The degree of diet variation $E = 0.87$ result was statistically highly significant ($p < 0.001$) and indicated a high variation among diets of individuals. The NODF = 22.4 metric indicated a moderate but significant ($p < 0.001$) nestedness in the network. At the same time, a high and significant modularity (Q = 0.58; $p < 0.001$) was present, and seven modules were identified (Figure 6).

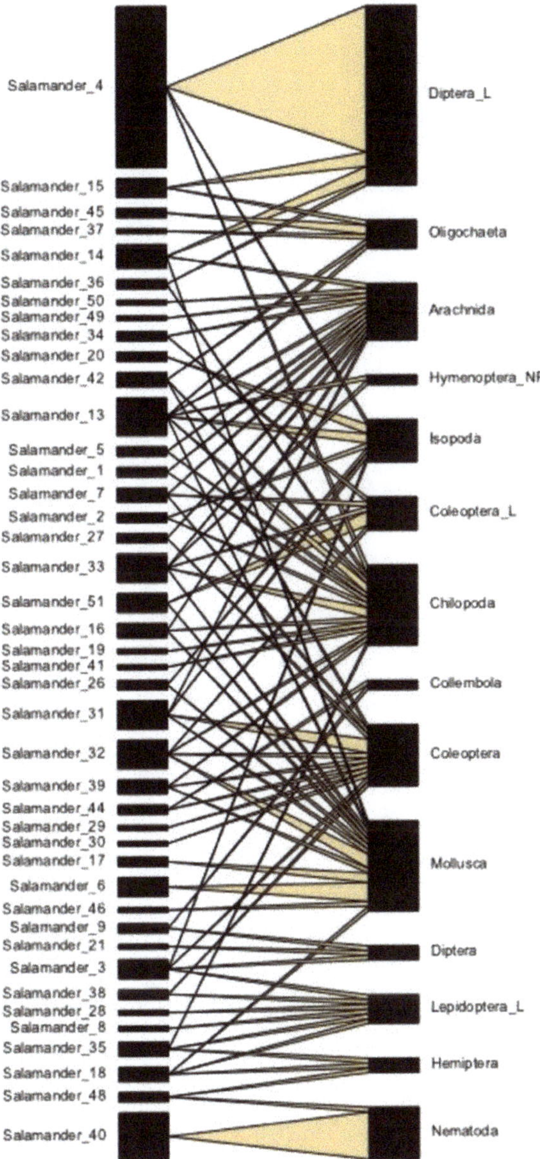

Figure 5. Bipartite network of salamanders and resources. Edge width represents frequency of use. The suffix L indicates the larval stage of the taxa; the suffix NF (not flying) indicates the not winged taxa or the not winged stage of a taxon.

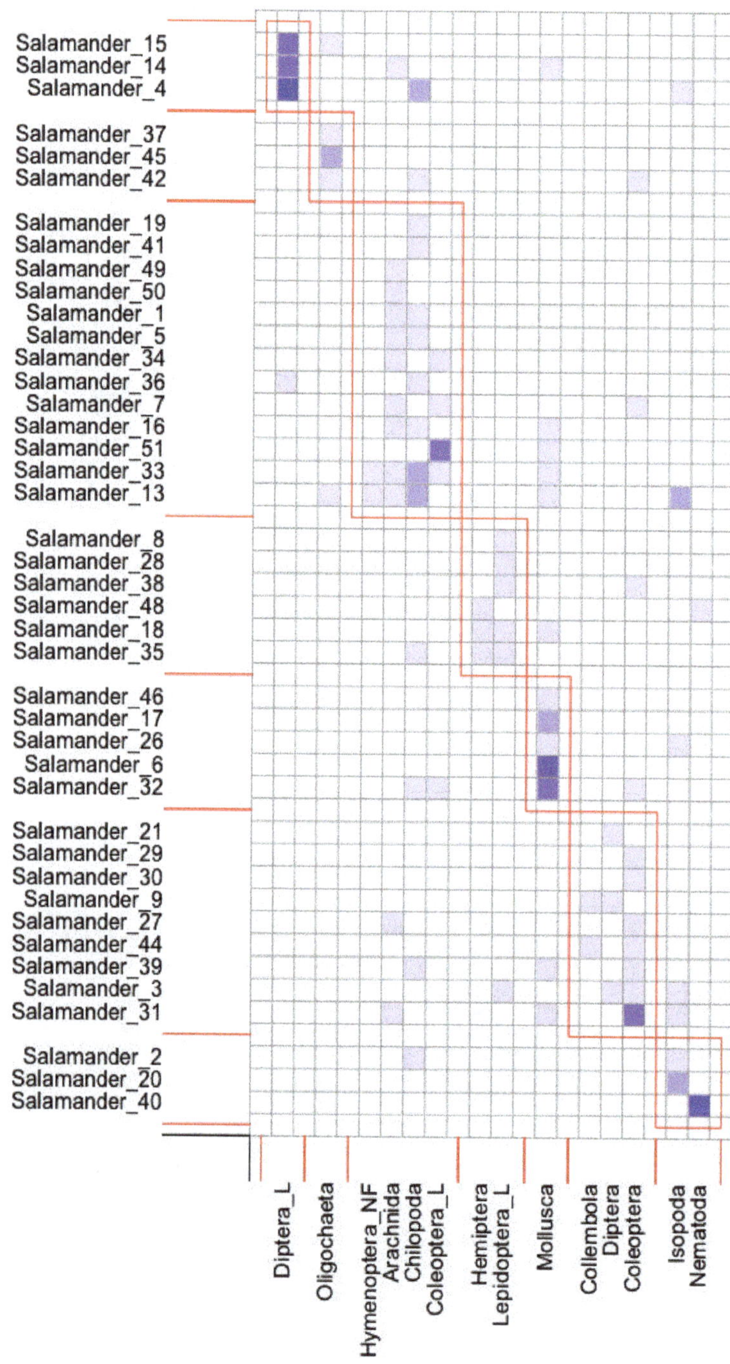

Figure 6. Adjacency matrix, with individual salamanders as rows and prey types as columns. Cell color represents frequency of use, red boxes indicate modules. The suffix L indicates the larval stage of the taxa; the suffix NF (not flying) indicates the not winged taxa or the not winged stage of a taxon.

4. Discussion

4.1. Prey Availability

Data about potential trophic niche highlighted rather low food availability, particularly in comparison with other studies about trophic strategy of other Italian salamanders carried out in similar periods with a similar number of pitfall traps [15,16,18]. These differences are not surprising due to the high-altitude alpine environment. Indeed, several studies confirm that conditions in mountains affect the species richness, composition, and density of different inver tebrates (e.g., [58–60]). Consequently, the low density of soil macroinvertebrates leads to a highly reduced amount of prey found in salamander's stomachs, in comparison with other species of terrestrial or semiterrestrial salamanders (e.g., ranging from 4.37 to 13.09 prey/stomach, [15]). The limited number of food items in the salamander stomachs may be due not only to the low prey availability, but also by the dimensional selectivity related to morphological characteristics of the predators and their predation strategy (see Section 4.2. and Section 4.3). ANOSIM and the diversity permutation test showed significant differences between total and nocturnal trophic availability, which strongly suggest that the actual daily activity pattern of the studied species must be taken into consideration because analyses that do not consider this factor can lead to misleading conclusions.

4.2. Realized Trophic Niche

There were no differences between males and females, according to previous studies [15,61,62]. The graphical Costello's method modified by Amundsen [48,49] identified a generalized trophic strategy in which Coloeptera and Mollusca represent the most valuable taxa of the diet (Figure 1). Diptera larvae is the only prey's category located in the upper-left quadrant of graph, indicating specialization. However, a low number of individuals preyed on this taxon ($n = 4$), and for only one of them, Diptera larvae represented 82% (31 of 38) of the diet. In this case, the high level of individual specialization, as exhibited by a single salamander, plays a marginal role in the overall picture of the population trophic strategy. The graphical plot showed that Alpine salamander avoided the use of small-sized taxa like Collembola, Diptera, and Formicidae, even if they are the dominant ones in the environment. The avoidance of small-sized taxa can be explained by considering the low number of preys/stomach (2.78) and the big size of this salamander. In fact, consumption of preys in amphibians is largely bound by morphological and physiological characteristics [63]. Furthermore, for Urodela, a positive correlation between dimension of the predator and prey has long been known [64]. Probably, predation strategy, not particularly refined and efficient to catch skillful prey, can also represent an important limit to usage of small-sized preys. Trophic strategy of *S. atra* pointed out by our study only partially agrees with observational information obtained in another Italian population [23], which exhibited a preference for Mollusca and Oligochaeta, followed by Orthoptera, Homoptera, Dermaptera, Isopoda, and Arachnida. Unfortunately, we could not perform an accurate comparison due to an absence of numerical data. Conversely, the study carried out by Fachbach et al. [30] on two German populations, although it did not report data for inferences on trophic strategy and selectivity, it provided numerical data on ingested preys. From the comparison of our data to those of the German populations (Figure 2), our population differed in the proportion of the prey taxa while the two German populations were more similar to each other (Figure 2). However, the overall diversity index of preyed taxa among the three populations was extremely similar. The most used categories in Fachbach and coauthors' study [30] were Coleoptera, Arachnida, and Coleoptera larvae, while few taxa seem to characterize just one population (e.g., Collembola in Population 2). In contrast, our study highlighted a preference for Diptera larvae, Mollusca, Chilopoda, and Coleoptera. Although the sampling methods to obtain information on preyed taxa were different (Fachbach and coauthors euthanized the salamanders), both methods collected preys directly by the stomach and may be considered comparable. Sample sizes were also comparable, although our salamanders belonged to the same population while the German ones were sampled from two different and distant populations. Consequently, differences among these populations in the ratio of preyed taxa may depend either on ecological variables in different environments (e.g., food availability) or on different behavioral traits at interpopulation level, or on both. The two German salamander populations [30] seem to occur in similar habitat (1200 m and 1600–1800 m a.s.l., calcareous geological substrate, open habitats mixed with coniferous woodland). Unfortunately, Fachbach et al. [30] reported only simple frequencies of preyed taxa for each population, but he did not report the number and type of prey per salamander and did not include trophic availability. Consequently, comparisons of electivity index or Amundsen method are inapplicable. However, the similar low average number of prey/salamander (2.78 from our study and 2.58 from Fachbach et al. [30]) suggest that the low number of ingested preys may be regarded as a characteristic trait of this species rather than that of a given population.

4.3. Trophic Selectivity

Definition of potential trophic niche is an indispensable tool to study trophic strategy and selectivity. Therefore, it is crucial to obtain data that are a good representation of the real trophic availability accessible to a given species. Dietary studies on other salamanders tried to achieve this target by using different types of traps that can collect samples from different habitats, like pitfall, sticky traps, and Berlese–Tullgren extractor [15,16]. Such methods are really effective to obtain data about the potential trophic niche, but should be used considering the activity period of the target species. Our results, for the first time, demonstrated that taking into account the temporal activity pattern of a predator may lead to a different interpretation of the trophic strategy. In literature, all studies about trophic strategy of salamanders perform arthropod sampling continuously for several days without considering the daily activity (and foraging) pattern of the studied species. However, we demonstrated that significant differences between total and nocturnal trophic availability could lead to different or even opposite interpretation of selectivity, which may also involve the most relevant taxa in the realized trophic niche. Electivity index, calculated with nocturnal trophic availability, shows a positive selection for Coleoptera and Chilopoda, highlighting the important role in summer's trophic strategy of these taxa. Mollusca and Arachnida were also identified as relevant preys. The first was selected almost in relation to their abundance in the environment (they are just above the significance threshold, Figure 3), the second were exposed to a counter selection. Generally, relativized electivity index showed a negative selection for Collembola, Formicidae, and Diptera, which are, however, the most abundant in trophic availability. Therefore, selectivity results were essentially in agreement with tropic strategy defined by Amundsen's method. As reported in Section 4.2, selection of prey type could be explained by dimension and hunting strategy of the Alpine salamander. Small number of preys per stomach could indicate a dimensional selectivity; for a large species such as *Salamandra atra*, foraging on few but large preys seems more worthwhile than eating many little ones, considering that a predator tends to maximize the energy intake consuming preys that give the best energy gain per time unit [65,66]. Given that selection seems to operate more on the prey's size than on taxonomic categories, it will be useful to analyze trophic selectivity of Alpine salamander using volumetric classes that represent a possible measure of energy-intake [15], which can show a more detailed overview of trophic strategy of *Salamandra atra*.

4.4. Interindividual Diet Variation

The presence of interindividual diet variation in *Salamandra atra* is consistent with what was observed in other salamander species (e.g., [16,18]). Patterns of interindividual diet variation can be explained by optimal diet theory (ODT) [67,68], since individuals may have different traits affecting their ability to capture or handle different prey types and they will rank prey preferences accordingly. In this context, ODT theory accounts for three distinct patterns that may cause individual specialization [69]. Individuals may have different rank preferences and therefore specialize on different prey types. When competition is present, individuals should include new resources in their diet and increase overlap; we call it the distinct preferences model. Alternatively, under the competitive refuge model, individuals may share the first ranked prey and have different rankings for the less preferred ones. It implies that, under intraspecific competition, individuals will broaden their trophic niche, including different prey types according to their ranking preferences. The third model, the shared preferences model, assumes that individuals have identical preferences but differ in the willingness to include lower ranked prey in their diet. Consequently, concerning network analysis, the shared preferences model suggests the presence of significant nestedness [18] under intraspecific competition, but no modularity could be observed. Modularity in turn, should be present in case of the distinct preferences model, when resources are unlimited and competition is low, or by contrast, in the competitive refuge model when resources are limited and competition is high [53]. In our case study, the observed emergence of a significant modular network may be explained by both the competitive refuge model and the distinct preferences model. If the available resources are low, intraspecific competition is present and the preferred shared resource becomes scarce or less profitable, driving the individuals to the inclusion of different prey types, and to the emergence of several modules. The observed modules, however, could also arise from a distinct preferences model, where resources are unlimited and individuals, released by competition, specialize on their preferred prey items.

5. Conclusions

The diet of Salamandra atra, a fully terrestrial salamander, is investigated in depth for the first time, although during a short timeframe in the first half of August (i.e., about in the midseason of activity [23]). It showed a generalist trophic strategy; not all prey were consumed proportionally to their environmental availability. Considering diet variation at the individual level, the competitive refuge model and the distinct preferences model, within optimal diet theory, seem to equally fit our results. However, given the low trophic availability, the former is more supported. As a general rule, if the sampling period was short, it may not describe exhaustively the feeding habits of a given population. Although it is true that cross-sectional studies may overestimate individual specialization, in particular when prey distribution reflects a patchy environment [13,17], but our sampling was performed in a relatively small and homogenous area and this should reduce the bias. Furthermore, cross-sectional design represents the majority of data-type used in the analysis of interindividual diet variation (see reviews by [17,18]). Considering the lack of knowledge about trophic habits of Alpine salamander, the study of the trophic strategy is the first crucial step to understand the ecological role of this species. Further research will be necessary to investigate any possible variations in diet during the whole activity season and the functional relationship existing between predator and its prey, in terms of relative abundance and biomass. Finally, as a methodological contribution, we highlighted that considering the daily activity pattern of a species is a fundamental requirement for correct data collection and resulting interpretation. Our study was performed on a population of the nominal subspecies. Further investigations on the others, three subspecies with two of them extremely localized and Italian endemic [70,71], may elucidate any feeding differences among these taxa.

Author Contributions: Conceptualization, A.R; Methodology and sampling, L.R. and A.R.; Formal analysis, L.R., A.R., and A.C.; Data curation, L.R. and A.R.; Writing—original draft preparation, A.R., L.R., and A.C.; Writing—review and editing, A.R., L.R., A.C., P.P, S.L., and G.M.; Coordinator of the project R2000 monitoring, P.P. All authors have read and agreed to the published version of the manuscript.

Funding: This research received no external funding.

Acknowledgments: This paper is the result of the master thesis of Luca Roner. The study received funding within the agreement 2017–2019 between MUSE and CNR-ISAFOM. Experimental protocols were approved by the Italian Ministry of Environment (authorization PNM-EU-2017-005370). This study was supported by Servizio Aree Protette e Sviluppo Sostenibile PAT (Piano di monitoraggio fauna vertebrata Direttiva Habitat; Progetto LIFE11/NAT/IT/000187 "TEN"–Trentino Ecological Network) and by the Paneveggio–Pale di San Martino Natural Park, which also provided logistic facilities close to the sampling site. In particular, we are grateful to the Technical Coordinator Piergiovanni Partel, and to the Director Vittorio Ducoli.Thanks to Michele Chiacchio, Francesco Maria Romano, and the park guard Gilberto Volcan, who contributed to the field sampling.

Conflicts of Interest: There are no conflicts of interest associated with this publication and there was no significant financial support for this work that could have influenced its outcome.

References

1. Davic, R.D.; Welsh, H.H., Jr. On the ecological roles of salamanders. *Ann. Rev. Ecol. Evol. System.* **2004**, *35*, 405–434. [CrossRef]
2. Hairston, N.G. The local distribution and ecology of the plethodontid salamanders of the southern Appalachians. *Ecol. Monogr.* **1949**, *19*, 49–73. [CrossRef]
3. Petranka, J.W. *Salamanders of the United States and Canada*; Smithsonian Institution Press: Washington, DC, USA, 1998; pp. 1–587.
4. Costa, A.; Crovetto, F.; Salvidio, S. European plethodontid salamanders on the forest floor: Local abundance is related to fine–scale environmental factors. *Herpetol. Conserv. Biol.* **2016**, *11*, 344–349.
5. Romano, A.; Anderle, M.; Forti, A.; Partel, P.; Pedrini, P. Population density, sex ratio and body size in a population of *Salamadra atra atra* on the Dolomites. *Acta Herpetol.* **2018**, *13*, 195–199.
6. Burton, T.M.; Likens, G.E. Energy flow and nutrient cycling in salamander populations in the Hubbard Brook Experimental Forest, New Hampshire. *Ecology* **1975**, *56*, 1068–1080. [CrossRef]
7. Best, M.L.; Welsh, H.H., Jr. The trophic role of a forest salamander: Impacts on invertebrates, leaf litter retention, and the humification process. *Ecosphere* **2014**, *5*, 1–19. [CrossRef]
8. Barton, K.A.; Zalewski, A. Winter severity limits red fox populations in Eurasia. *Glob. Ecol. Biogeogr.* **2007**, *16*, 281–289. [CrossRef]
9. Isbell, L.A. Sudden short–term increase in mortality of vervet monkeys (*Cercopithecus aethiops*) due to leopard predation in Amboseli National Park, Kenya. *Am. J. Primatol.* **1990**, *21*, 41–52. [CrossRef]
10. Berger, L.; Speare, R.; Thomas, A.; Hyatt, A. Mucocutaneous fungal disease in tadpoles of *Bufo marinus* in Australia. *J. Herpetol.* **2001**, *35*, 330–335. [CrossRef]
11. Chapman, C.A.; Wasserman, M.D.; Gillespie, T.R.; Speirs, M.L.; Lawes, M.J.; Saj, T.L.; Ziegler, T.E. Do food availability, parasitism, and stress have synergistic effects on red colobus populations living in forest fragments? *Am. J. Phys Anthropol.* **2006**, *131*, 525–534. [CrossRef]
12. Hanya, G.; Chapman, C.A. Linking feeding ecology and population abundance: A review of food resource limitation on primates. *Ecol. Res.* **2013**, *28*, 183–190. [CrossRef]
13. Bolnick, D.I.; Yang, L.H.; Fordyce, J.A.; Davis, J.M.; Svanback, R. Measuring individual–level resource specialization. *Ecology* **2002**, *83*, 2936–2941. [CrossRef]
14. Wells, K.D. *The Ecology and Behavior of Amphibians*; Chicago University Press: Chicago, IL, USA, 2007.
15. Salvidio, S.; Romano, A.; Oneto, F.; Ottonello, D.; Michelon, R. Different season, different strategies: Feeding ecology of two syntopic forest dwelling salamanders. *Acta. Oecol.* **2012**, *43*, 42–50.
16. Costa, A.; Salvidio, S.; Posillico, M.; Matteucci, G.; De Cinti, B.; Romano, A. Generalisation within specialization: Inter–individual diet variation in the only specialized salamander in the world. *Sci. Rep.* **2015**, *5*, 13260. [CrossRef] [PubMed]
17. Bolnick, D.I.; Svanback, R.; Fordyce, J.A.; Yang, L.H.; Davis, J.M.; Hulsey, C.D.; Forister, M.L. The ecology of individuals: Incidence and implications of individual specialization. *Am. Nat.* **2003**, *161*, 1–28. [CrossRef]
18. Araújo, M.S.; Martins, E.G.; Cruz, L.D.; Fernandes, F.R.; Linhares, A.X.; Dos Reis, S.F.; Guimaraes, P.R., Jr. Nested diets: A novel pattern of individual-level resource use. *Oikos* **2010**, *119*, 81–88. [CrossRef]
19. Salvidio, S.; Oneto, F.; Ottonello, D.; Costa, A.; Romano, A. Trophic specialization at the individual level in a terrestrial generalist salamander. *Can. J. Zool.* **2015**, *93*, 79–83. [CrossRef]
20. Salvidio, S.; Costa, A.; Crovetto, F. Individual trophic specialisation in the Alpine newt increases with increasing resource diversity. *Ann. Zool. Fenn.* **2019**, *56*, 17–24. [CrossRef]
21. Rosa, G.; Costa, A.; Salvidio, S. Trophic strategies of two newt populations living in contrasting habitats on a Mediterranean island. *Ethol. Ecol. Evol.* **2020**, *32*, 96–106. [CrossRef]
22. Costa, A.; Baroni, D.; Romano, A.; Salvidio, S.; Lötters, S. Individual diet variation in *Salamandra salamandra* larvae in a Mediterranean stream (Amphibia: Caudata). *Salamandra* **2017**, *53*, 148–152.
23. Lanza, B.; Andreone, F.; Bologna, M.A.; Corti, C.; Razzetti, E. *Fauna d'Italia, vol. XLII, Amphibia*; Calderini Edition: Bologna, Italy, 2007; pp. 197–211.
24. Sillero, N.; Campos, J.; Bonardi, A.; Corti, C.; Creemers, R.; Crochet, P.A.; Isailovic, J.C.; Denoël, M.; Ficetola, G.F.; Gonçalves, J.; et al. Updated distribution and biogeography of amphibians and reptiles of Europe. *Amphib. Reptil.* **2014**, *35*, 1–31. [CrossRef]

25. Bedriaga, J. von. *Die Lurchfauna Europa's II. Urdodela. Schwanzlurche.* Univ. typ. Moscow. Available online: http://www.europeana.eu/en/item/08711/item_23042?fbclid=IwAR368Ijnp_DRjaXfZmkOrOKIrDYic6VsgrfQ6EhAjySPxW9OcsBMZuga4dE (accessed on 16 May 2020).

26. Freytag, G.E. *Feuersalamander und Alpensalamander. Die Neue Brehm-Bücherei*; A. Ziemsen Verlag: Wittenberg Lutherstadt, Germany, 1955; p. 142.

27. Klewen, R.F. Untersuchungen zur Verbreitung, Öko-Ethologie und innerartlichen Gliederung von *Salamandra atra* (Laurenti 1768). Ph.D. Thesis, Universität Köln, Köln, Germany, 1986.

28. Klewen, R.F. *Die Landsalamander Europas 1: Die Gattungen Salamandra und Mertensiella*; A. Ziemsen Verlag: Wittenberg Lutherstadt, Germany, 1988.

29. Kuzmin, S.L. Feeding ecology of *Salamandra* and *Mertensiella*: A review of data and ontogenetic evolutionary trends. *Mertensiella* **1994**, *4*, 271–286.

30. Fachbach, G.; Kolossau, I.; Ortner, A.; Zur Ernährungsbiologie von Salamandra, s. salamandra und *Salamandra atra*. *Salamandra* **1975**, *11*, 136–144.

31. Scheller, H.V. Pitfall trapping as the basis for studying ground beetle predation in spring barley. *Tidsskr. Planteavl.* **1984**, *88*, 317–324.

32. Woodcock, B.A. Pitfall trapping in ecological studies. In *Insect Sampling in Forest Ecosystems*; Leather, S.R., Ed.; Blackwell Publishing Co.: New Jersey, NJ, USA, 2005.

33. Southwood, T.R.E.; Henderson, P.A. *Ecological Methods*, 3rd ed.; Wiley-Blackwell: Oxford, UK, 2000; p. 575.

34. Ausden, M. Invertebrates. In *Ecological Census Techniques: A Handbook*; Sutherland, W.J., Ed.; Cambridge University Press: Cambridge, UK, 1996.

35. Duellman, W.E.; Trueb, L. *Biology of Amphibians*; Johns Hopkins Univ. Press: Baltimore, MD, USA, 1994; p. 670.

36. Solé, M.; Rödder, D. Dietary assessments of adult amphibians. In *Amphibian Ecology and Conservation: A Handbook of Techniques*; Dodd, J., Ed.; Oxford University Press: Oxford, UK, 2010.

37. Fraser, D.F. Coexistence of salamanders in the genus Plethodon: A variation of the Santa Rosalia theme. *Ecology* **1976**, *55*, 238–251. [CrossRef]

38. Salvidio, S. Diet and food utilization in a rockface population of *Speleomantes ambrosii* (Amphibia, Caudata, Plethodontidae). *Vie et Milieu* **1992**, *42*, 35–39.

39. Schabetsberger, R. Gastric evacuation rates of adult and larval Alpine newts (*Triturus alpestris*) under laboratory and field conditions. *Freshw. Biol.* **1994**, *31*, 143–151. [CrossRef]

40. Caldwell, J.P. The evolution of myrmecophagy and its correlates in poison frogs (Family Dendrobatidae). *J. Zool.* **1996**, *240*, 75–101. [CrossRef]

41. Solé, M.; Beckmann, O.; Pelz, B.; Kwet, A.; Engels, W. Stomach–flushing for diet analysis in anurans: An improved protocol evaluated in a case study in Araucaria forests, southern Brazil. *Stud. Neotrop. Fauna Environ.* **2005**, *40*, 23–28. [CrossRef]

42. Clarke, K.R. Non–parametric multivariate analyses of changes in community structure. *Aust. J. Ecol.* **1993**, *18*, 117–143. [CrossRef]

43. Magurran, A.E. *Measuring Biological Diversity*; Blackwell Publishing: Oxford, UK, 2004; p. 256.

44. Wu, Z.-J.; Li, Y.; Wang, Y. A comparison of stomach flush and stomach dissection in diet analysis of four frog species. *Acta Zool. Sin.* **2007**, *53*, 364–372.

45. Hammer, O.; Harper, D.A.T.; Ryan, P.D. PAST: Paleontological statistics software package for education and data analysis. *Paleontol. Electron.* **2001**, *4*, 1–9.

46. Vanderploeg, H.A.; Scavia, D. Calculation and use of selectivity coefficients of feeding: Zooplankton grazing. *Ecol. Mod.* **1979**, *7*, 135–149. [CrossRef]

47. Lechowicz, M.J. The sampling characteristics of electivity indices. *Oecologia* **1982**, *52*, 22–30. [CrossRef] [PubMed]

48. Ramos–Jiliberto, R.; Valdovinos, F.S.; Arias, J.; Alcaraz, C.; García–Berthou, E. A network–based approach to the analysis of ontogenetic diet shifts: An example with an endangered, small–sized fish. *Ecol. Compl.* **2011**, *8*, 123–129. [CrossRef]

49. Costello, M.J. Predator feeding strategy and prey importance: A new graphical analysis. *J. Fish Biol.* **1990**, *36*, 261–263. [CrossRef]

50. Amundsen, P.A.; Gabler, H.M.; Staldvik, F.J. A new approach to graphical analysis of feeding strategy from stomach contents data—Modification of the Costello (1990) method. *J. Fish Biol.* **1996**, *48*, 607–614.

51. Araújo, M.S.; Guimarães, P.R., Jr.; Svanbäck, R.; Pinheiro, A.; Guimarães, P.; Reis, S.F.D.; Bolnick, D.I. Network analysis reveals contrasting effects of intraspecific competition on individual vs. population diets. *Ecology* **2008**, *89*, 1981–1993. [CrossRef]

52. Pires, M.M.; Guimarães, P.R., Jr.; Araújo, M.S.; Giaretta, A.A.; Costa, J.C.L.; Dos Reis, S.F. The nested assembly of individual-resource networks. *J. Anim. Ecol.* **2011**, *80*, 896–903. [CrossRef]

53. Tinker, M.T.; Guimarães, P.R., Jr.; Novak, M.; Marquitt, F.M.D.; Bodkin, J.L.; Staedler, M.; Bentall, G.; Estes, J.A. Structure and mechanism of diet specialisation: Testing models of individual variation in resource use with sea otters. *Ecol. Lett.* **2012**, *15*, 475–483. [CrossRef]

54. Santamaría, S.; Enoksen, C.A.; Olesen, J.M.; Tavecchia, G.; Rotger, A.; Igual, J.M.; Traveset, A. Diet composition of the lizard *Podarcis lilfordi* (Lacertidae) on 2 small islands: An individual–resource network approach. *Curr. Zool.* **2020**, *66*, 39–49. [CrossRef]

55. Blüthgen, N.; Menzel, F.; Blüthgen, N. Measuring specialization in species interaction networks. *BMC Ecol.* **2006**, *6*, 9. [CrossRef] [PubMed]

56. Almeida–Neto, M.; Guimaraes, P.; Guimaraes, P.R., Jr.; Loyola, R.D.; Ulrich, W. A consistent metric for nestedness analysis in ecological systems: Reconciling concept and quantification. *Oikos* **2008**, *117*, 1227–1239. [CrossRef]

57. Beckett, S.J. Improved community detection in weighted bipartite networks. *R. Soc. Open Sci.* **2016**, *3*, 140536. [CrossRef] [PubMed]

58. Aubry, M.P.; Berggren, W.A.; van Couvering, J.; McGowran, B.; Pillans, B.; Hilgen, F.J. Quaternary: Status, rank, definition, survival. *Episodes* **2005**, *28*, 118–120. [CrossRef] [PubMed]

59. Sfenthourakis, S.; Anastasiou, I.; Strutenschi, T. The terrestrial isopod diversity of Mt. Panachaiko (Pelopon–nisos, Greece). *E. J. Soil Biol.* **2005**, *41*, 91–98. [CrossRef]

60. Lessard, S.J.; Rivas, D.A.; Stephenson, E.J.; Yaspelkis, B.B., III; Koch, L.G.; Britton, S.L.; Hawley, J.A. Exercise training reverses impaired skeletal muscle metabolism induced by artificial selection for low aerobic capacity. *Am. J. Physiol. Regul. Integr. Comp. Physiol.* **2011**, *300*, R175–R182. [CrossRef]

61. Lamb, T. The influence of sex and breeding condition on microhabitat selection and diet in the Pig Frog *Rana grylio*. *Am. Mid. Nat.* **1984**, *111*, 311–318. [CrossRef]

62. Romano, A.; Salvidio, S.; Palozzi, R.; Sbordoni, V. Diet of the newt, *Triturus carnifex* (Laurenti, 1768), in a flooded karstic sinkhole ("Pozzo del Merro", Central Italy). *J. Cave Karst Stud.* **2012**, *74*, 271–277. [CrossRef]

63. Duellman, W.E.; Trueb, L. *Biology of Amphibians*; Mc Graw-Hill Book Company: New York, NY, USA, 1986.

64. Whitaker, J.O., Jr.; Rubin, D.C. Food habits of Plethodon jordani metcalfi and Plethodon jordani shermani from North Carolina. *Herpetologica* **1971**, *27*, 81–86.

65. Krebs, J.R. Optimal foraging: Decision rules for predators. In *Behavioural Ecology: An Evolutionary Approach*; Krebs, J.R., Davies, N.B., Eds.; Backwell Scientific Publications: Oxford, UK, 1978; pp. 23–63.

66. Stephens, D.W. *Foraging Theory*; Princeton University Press: Princeton, NJ, USA, 1986.

67. Schoener, T.W. Theory of feeding strategies. *Annu. Rev. Ecol. Syst.* **1971**, *2*, 369–404. [CrossRef]

68. Pulliam, H.R. On the theory of optimal diets. *Am. Nat.* **1974**, *108*, 59–75. [CrossRef]

69. Svanbäck, R.; Bolnick, D.I. Intraspecific competition affects the strength of individual specialization: An optimal diet theory method. *Evol. Ecol. Res.* **2005**, *7*, 993–1012.

70. Speybroeck, J.; Beukema, W.; Bok, B.; Van Der Voort, J. *Field Guide to the Amphibians and Reptiles of Britain and Europe*; Bloomsbury Natural History: London, UK, 2016.

71. Razpet, A.; Šunje, E.; Kalamujić, B.; Tulić, U.; Pojskić, N.; Stanković, D.; Krizmanić5, I.; Marić, S. Genetic differentiation and population dynamics of Alpine salamanders (*Salamandra atra*, Laurenti 1768) in Southeastern Alps and Dinarides. *Herpetol. J.* **2016**, *26*, 111–119.

Article

Feeding Strategies of Co-occurring Newt Species across Different Conditions of Syntopy: A Test of the "Within-Population Niche Variation" Hypothesis

Jennifer Mirabasso [1], Alessandra M. Bissattini [1], Marco A. Bologna [1], Luca Luiselli [2], Luca Stellati [1] and Leonardo Vignoli [1,*]

1   Department of Sciences, Roma Tre University, Viale Marconi, 446, 00146 Rome, Italy; jennifer.mirabasso@hotmail.it (J.M.); alessandramaria.bissattini@uniroma3.it (A.M.B.); marcoalberto.bologna@uniroma3.it (M.A.B.); luca.stellati@uniroma3.it (L.S.)
2   IDECC, Institute for Development, Ecology, Conservation and Cooperation, via G. Tomasi di Lampedusa 33, 00144 Rome, Italy; lucamlu@tin.it
*   Correspondence: leonardo.vignoli@uniroma3.it; Tel.: +39-06-57336411; Fax: +39-06-57336321

Received: 15 April 2020; Accepted: 5 May 2020; Published: 7 May 2020

Abstract: Intraspecific trait variation in generalist animals is widespread in nature, yet its effects on community ecology are not well understood. Newts are considered opportunistic feeders that may co-occur in different syntopic conditions and represent an excellent model for studying the role of individual feeding specialization in shaping the population dietary strategy. Here, we investigated the diet of three newt species from central Italy occurring in artificial habitats in different coexistence conditions to test the predictions of the niche width (NW) variation hypotheses. Population NW did not vary among species and between presence and absence of coexisting species. An overall positive relationship between individual specialization and population NW was observed. However, this pattern was disrupted by the condition of syntopy with newt populations showing an individual NW variation invariant with population NW in presence of coexisting species, whereas it was larger in populations occurring alone. The observed pattern of newt behavior was not consistent with any of the proposed scenarios. We found a consistent pattern with the degree of individual specialization being (1) size-dependent (specialized individuals increasing within larger sized species) and (2) assemblage-complexity-dependent (specialized individuals increasing in syntopic populations in comparison to singly populations).

Keywords: community ecology; *Triturus*; *Lissotriton*; coexisting species; trophic niche; niche width; niche variation hypothesis

---

1. Introduction

Urodeles are important elements of freshwater vertebrate communities either as prey or as predators [1–4] contributing to most of the total predator biomass in specific study areas [5]. They usually perform many "key ecological functions" [6] occurring at the top of the food chain [7–9] or, more frequently, at an intermediate level [10,11]. As predators of invertebrates and small vertebrates, they modulate energy pathways by decreasing the abundance of competitively dominant preys and consequently increasing taxa diversity in lower trophic levels [12]. Moreover, through their dual life cycle, they serve as connecting pathways for energy between aquatic and terrestrial environments affecting prey communities in both habitat types [13].

In general, newts are opportunistic feeders [14,15] consuming zooplankton, crustaceans, insects, fish, tadpoles, and even aerial or soil fauna fallen into the water during the aquatic phase [16–19], as well as isopods, diplopods, insects, and earthworms collected on the ground during the terrestrial phase [19–22]. Assemblages of newt species of differing body size are common throughout Europe, such as in Italy, where up to three species may co-occur in syntopy in the same aquatic site [16,21,23]. Because of their generalist feeding habits, and because they may co-occur in different sympatric conditions (see below), newts are excellent models for studying the role of individual feeding specialization in the "overall dietary niche" of a generalist feeder species/population. Indeed, it has been postulated that within-population niche variation can stabilize population and community dynamics [24], with environments having greater resource diversity favoring ecological diversity among consumers via disruptive selection or phenotypic and ecological plasticity or, as an alternative mechanism, that niche variation may be a consequence of neutral genomic diversity in more abundant populations [25].

Here, we investigated the diet of three newt species, the Italian newt (*Lissotriton italicus*), the Italian smooth newt (*Lissotriton vulgaris*), and the Italian crested newt (*Triturus carnifex*), in different co-occurring conditions in central Italy in order to test the predictions of the "within-population niche variation" theory across different species and co-presence conditions [26,27]. Except for a few theoretical studies [28,29], the "within-population niche variation" theory has rarely been tested before on natural assemblages of potentially competing species (but see [30,31]), but merely on populations of single species across several taxa (reviewed in [24,32]). Thus, our paper is one of the first to test the theory on an assemblage of species that have been considered in competition for the available resources in previous studies [33].

Specifically, the aims of the present study were to answer the following key questions:

(1) Does newt body size affect the tendency of individuals to partition their trophic niche by specializing the use of food resources towards distinct prey categories? Being newt gap-limited predators [11,33,34], we hypothesize that larger species feed on a large variety of prey items and thus have higher chance to differentiate the diet at the individual level to limit the potential impact of intraspecific competition.

(2a) Does the presence of taxonomically related species (congeneric or belonging to the same family) influence the feeding strategy of target newt populations in terms of trophic niche width and individual feeding behavior (i.e., degree of individual specialization)? We would expect that the co-occurrence of related species should promote differentiation of the trophic spectrum in conspecific individuals to mitigate the likely increased intra- and inter-specific competition [26]. (2b) Does the degree of syntopy (single vs. multispecies systems) affect newt body size that in turn may influence the feeding ecology of the species? We hypothesize that (a) large-bodied species outcompete smaller species through competition and intra guild predation, and (b) body size of smaller species are negatively affected by the occurrence of larger species. Moreover, we would expect that in multispecies conditions, the smaller species would suffer from the co-occurrence of large-bodied species due to the negative effect of intra guild predation (i.e., competition and predation by the same antagonist). Predation might influence individual behavior by constraining individuals to forage in restricted areas based on their boldness (i.e., propensity to forage in the presence of risk; [35]), and thus affecting the magnitude of individual specialization if resources are patchy [24]. Since the effect of interspecific competition and predation on individual specialization remains controversial both in the theoretical and empirical literature [24,30], we would expect species specific response by the various newt species.

In order to answer to the above-mentioned key questions, we surveyed specific artificial aquatic habitats (i.e., wells) characterized by a circular shape, vertical walls, and high depth (up to 6 m), generally associated with traditional agriculture and cattle watering. Indeed, wells represent ideal scale-effective systems to investigate interspecific interactions and community composition being characterized by a higher stability and a more simplified structure (e.g., limited volume, closed physical boundaries, simple and consistent vegetation structure and resource availability, absence of fish predators) in comparison to natural aquatic sites [36]. Moreover, at the study area, these habitats (i) are widespread, (ii) have been in place for a time long enough to enable the establishment of stable communities, (iii) host one up to three newt species syntopically, and (iv) are consistent in shape and size, thus, representing a self-set replicated experimental system [9,36].

2. Materials and Methods

2.1. Study Species

Lissotriton italicus (Peracca, 1898) and *L. vulgaris* (Linnaeus, 1758) are endemic to the southern Italian peninsula and widespread throughout Europe, respectively [22], whereas *T. carnifex* (Laurenti, 1768) is distributed through the Italian and northern Balkan peninsulas [22]. All the study species occur in natural permanent and temporary aquatic sites with stagnant or semi-flowing water, but they can also colonize artificial aquatic sites (tanks, drinking troughs, reservoirs, and wells) during the breeding season [22]. They are mostly threatened by fragmentation and loss of wetlands, pollution of aquatic habitats, and the introduction of alien fish [22,37,38]. However, only the Italian crested newt is listed on Appendix II of the Bern Convention and on Annexes II and IV of the Habitats Directive (92/43/EEC), while *L. italicus* is listed in the Annex IV.

Triturus carnifex attains the largest body size of any Italian newt (with females measuring 120–180 mm and males 100–150 mm in total length) and it often co-occurs with *L. vulgaris* (60–110 mm for males and females) and *L. italicus* (with a total length of up to 80 mm for the larger females, this species is considered the smallest of the European newts) [22]. Contrarily, *L. italicus* rarely lives in syntopy with *L. vulgaris* because of their mostly allopatric distribution [22]. *Triturus carnifex* and *L. vulgaris* are the species less dependent on an aquatic environment [22].

2.2. Study Area

Field work was carried out in the Aurunci Mountains, part of the Volsci range, constituted by the Lepini-Ausoni-Aurunci Mts. and forming a limestone chain parallel to Apennines and close to the Tyrrhenian Sea, located in the southern part of the Latium Region (central Italy). They are part of the "Monti Aurunci Regional Park" (surface: 20,000 ha) established in 1997 and characterized by a mosaic of complex and heterogeneous landscapes [39]. The vegetation includes Mediterranean scrubs (*Spartium junceum* L., *Myrtus communis* L., *Pistacia lentiscus* L., *Arbutus unedo* L., *Calluna vulgaris* L., *Erica spp.*) and woodlands (*Quercus ilex* L.) in the southern slope, whereas several arboreal species (*Ostrya carpinifolia* Scopoli, *Carpino orientalis* Miller, *Fraxinus ornus* L.) have colonized the northern slope [39,40]. At higher altitudes forests with presence of *Fagus sylvatica* L. are intermitted with grasslands [39,41].

The area considered in the present study is located between 41°27′ N and 41°18′ N latitude and 12°23′ E–13°45′ E longitude, with an extent of about 400 km^2 (Figure 1). A comprehensive survey of wells was conducted by locating them on topographical maps (Google Earth©; Figure 1), and 17 artificial aquatic sites were located and geo-referred in the study area.

Wells are characterized by a relatively thick aquatic vegetation, mainly *Potamogeton* spp. Overall, the macrophytic flora is scant and limited to a few small patches of riparian vegetation and algae. Each aquatic site was sampled at least once during the aquatic phase of newt species (March–July; [22]) (Figure 2).

At the study area, wells are important elements of the landscape as they are used to help traditional husbandry and agriculture because of the scarcity of natural aquatic systems due to the widespread karst phenomena [42]. The three newt species studied in this paper do occur in wells either alone or in syntopy, but *L. vulgaris* rarely occurs alone or in exclusive co-presence with *L. italicus*. In the present paper, new species foraging in the same well, representing a physically closed habitat isolated from the other aquatic sites, were considered as syntopic. On the other hand, species that live in areas with several wetlands, whose borders are arbitrarily defined, often at a larger scale than that perceived by the species, i.e., a lake or river floodplain [8], a forest [43], a mountainous system [44], or a protected area [45], can be defined as sympatric, and thus, by potentially exploiting distinct habitats for reproduction and/or feeding, may not interact at all [19].

Figure 1. Map illustrating the distribution of the study wells and the newt species with indication of species coexistence. The numbers on the map refer to the well ID (see Table S1) (Google Earth, earth.google.com/web/).

Figure 2. A typical well used as water reservoir in the study area. All the sampled wells, characterized by this same structure consisting of vertical stony walls with the upper margin just one small step above the ground, are easily colonized by newts.

2.3. Sampling

Newts were collected (Italian Ministry of the Environment gave LV the permit [0001255/PNM] to conduct stomach flushing and manipulate amphibians) by daytime (from 9:00 a.m. to 6:00 p.m.) during the breeding season (March–July 2019) [22]. Individuals were visually located and captured when surfacing to breathe by using a long-handled dip net (3.5 m in length) from the shore. Immediately after capture, newts were marked by a photograph of the ventral pattern to avoid pseudoreplication, measured (SVL = snout-vent length to the nearest mm) and sexed based on secondary sexual characters [22]. We sampled each well from one to three times to gather enough individuals ($n > 7$) for each population. Unfortunately, we did not have sufficient recapture data to provide good-enough estimates on density. All the analyses were carried out on adult newts. Stomach contents were collected by stomach flushing [46], individually stored in vials containing 70% ethanol solution and analyzed in the laboratory. Collected newts were temporarily housed in tanks filled with water for approximately two hours after flushing to verify their return to normal activity, and then released at the same point of capture. No mortality was observed during or after stomach flushing. Taxonomic identification of stomach contents was made using a stereomicroscope (Olympus SZX 12. Range of magnification 9–55×). Food items were classified to the lowest taxonomic level possible (usually order or family). Prey composition in newt diet was quantified by estimating the number and occurrence of each food item. The ingestion of plants and minerals was considered accidental and not included in further analyses. We did not estimate prey availability in the environment due to logistic issues. However, it can be assumed that the study wells were homogenous enough in terms of size, structure, and aquatic habitat to limit the variability of trophic resource availability across sites. Although an overall positive effect of ecological opportunity on diet variation is reported for some taxa (i.e., higher prey diversity likely creates more opportunity for individual diet variation; [30,47]), it is very likely that in most habitats, ecological opportunity (i.e., prey availability) may not be limited for generalist predators such as newts, and thus, that it does not represent a sufficient constraint to refrain individual diets from diverging [15,19], so that it may not be biologically important for generalist species in diverse communities [30]. Moreover, the empirical estimate of prey availability should be considered a proxy of, and may diverge from, the actual consumer perception of ecological opportunity due to sampling methods and effort, thus leading to unreliable inferences [30].

2.4. Data Analyses

Only populations represented by $n > 7$ collected individuals were considered in the analyses. Biometric variables were log-transformed to meet the assumptions of parametric tests. Factorial analysis of variance (ANOVA) was performed to compare SVL among newt species and among study sites and to identify intersexual differences within and among species. ANOVA was also used to test differences in body size for *T. carnifex* and *L. italicus* when they occurred as a single species or in syntopy with other newt species. *Lissotriton vulgaris* was excluded from this latter analysis, since it was always found in syntopy with at least one of the other newt species.

The within-population variation in diet was assessed through the proportional similarity index (PS_i) [48,49] that compares the resource use distribution of an individual to that of its population in terms of overlap:

$$PS_i = 1 - 0.5 \sum \left| p_{ij} - q_j \right| = \sum \min\left(p_{ij}, q_j \right), \tag{1}$$

where p_{ij} is the frequency of diet category j in the individual i's diet, and q_j is the frequency of diet category j in the population as a whole. PS_i equal to q_j identified individuals specialized on a single diet item j, whereas PS_i values equal to 1 corresponded to conformer individuals (i.e., those consuming resource proportionally to population). The overall prevalence of individual specialization (IS) in the population was expressed by the average PS_i value:

$$IS = 1/N \sum PS_i, \tag{2}$$

IS varies between a value close to 0 (strong individual specialization) and 1 (no individual specialization) [50]. We then represented the population diet variation as $V = 1 - IS$, which ranges from 0 (all individuals use the full range of resources exploited by the population) to higher decimal values (individuals use only a subset of their population's diet spectrum) [26].

Population niche breadth was estimated by calculating the total niche width (TNW) of each group quantified by means of the Shannon–Weaver diversity index, following Roughgarden [51]. This index will yield a value of 0 when the entire population uses only a single category of prey, increasing with both the number of prey categories and the evenness with which they are used.

We regressed V on TNW for all the newt populations in order to explore how the relationship between niche width and individual specialization vary among species and in different condition of species co-occurrence. Because the evaluation and comparison (i.e., two populations or the same population at different points) of niche indices are affected by the limitation of arbitrary cut-offs [52], we compared the observed overlap values to an appropriate null model [48]. For each group of individuals, we first pooled all prey counts and determined the frequency of each prey category in the summed population diet. Each individual, observed to have consumed n number of prey items, was then randomly reassigned n items via multinomial sampling from the population diet frequencies. The null degree of IS and diet variation (V) was calculated once all individuals were assigned random diets. For each group, we carried out 999 such resampling estimates. We then regressed the mean resampled V against the observed TNW, to evaluate the null hypothesis that limited individual diet data also generate a positive relationship between these measures. To evaluate whether our observed trend can be explained by this null model alone, we used a General Linear Model (GLM) to test for a difference between the slopes of the observed and simulated V (data type) against TNW (significant interaction term between TNW and "data type" factor indicates that the observed and simulated slopes differ). The same analysis performed on each newt species was conducted on all the species pooled, thus considering all the newt individuals as a model organism irrespective of the species they belong to. This latter analysis allowed to test the overall effect of the degree of syntopy (single species vs. co-occurring species) on the relationship between V and TNW.

The disadvantage of using IS when analyzing different species at distinct locations is represented by the difficulty to interpret high or low values as, respectively, low or high degrees of individual specialization (i.e., arbitrary cut-offs). Indeed, the absolute values index does not unambiguously indicate to what extent the observed variations represent significant differences in individual diet width. The comparison of the observed degree of individual specialization to the distribution of a null population consisting of generalist (i.e., conformer) individuals allows to overcome this potential critical issue. Thus, for those newt species that occurred as single species or syntopic with other species, we used a GLM to test the effect of the species co-occurrence on individual trophic specialization using the percentage of "simulated Psi > observed Psi" as dependent variable.

Statistical tests were performed using STATISTICA software (version 8.0 for Windows) with alpha set at 5%. Both TNW and PS_i were calculated in IndSpec1.0 [48].

3. Results

3.1. Body Size

Overall, 562 adult newts were collected from 25 distinct populations with $n > 7$ (236 *T. carnifex*, 172 *L. vulgaris* and 154 *L. italicus*) (Table S2). Significant differences in SVL among species were found, with *T. carnifex* > *L. vulgaris* > *L. italicus* (Table 1; Figure 3A; for all post hoc comparisons, $p < 0.001$; refer to Table S2 for a synopsis of newt body size in all sampled sites). Body size differed among study sites and intersexual differences were found within each newt species, with females being larger than males (Table 1; Figure 3A). Both *T. carnifex* and *L. italicus* SVL varied significantly in the presence/absence of other newt species (effect of syntopy: Table 1; Figure 3B): *T. carnifex* was larger when the species occurred in syntopy with other newt species, whereas for *L. italicus* the opposite was true.

Table 1. Effect of site (random factor), species, sex, syntopy condition (yes/no), and interaction terms of fixed factors on newt body size (snout-vent length to the nearest mm; SVL) at the study sites. The effect of syntopy condition (°) was tested on *Lissotriton italicus* and *Triturus carnifex* only, because *Lissotriton vulgaris* was always found in co-occurrence with other species. Abbreviations: SS = sum of squares; MS = mean squares. Significant effects are highlighted in bold.

Effect	SS	DoF	MS	F	*p*
Intercept	130.300	1	130.300	15,218.71	<0.0001
Site	0.227	16	0.014	14.18	<0.0001
Species	6.605	2	3.302	3301.11	<0.0001
Sex	0.145	1	0.145	144.79	<0.0001
Species × Sex	0.048	2	0.024	23.99	<0.0001
Syntopy°	0.002	1	0.002	1.01	0.315
Species × Syntopy°	0.014	1	0.014	8.48	0.004
Error	0.540	540	0.001		

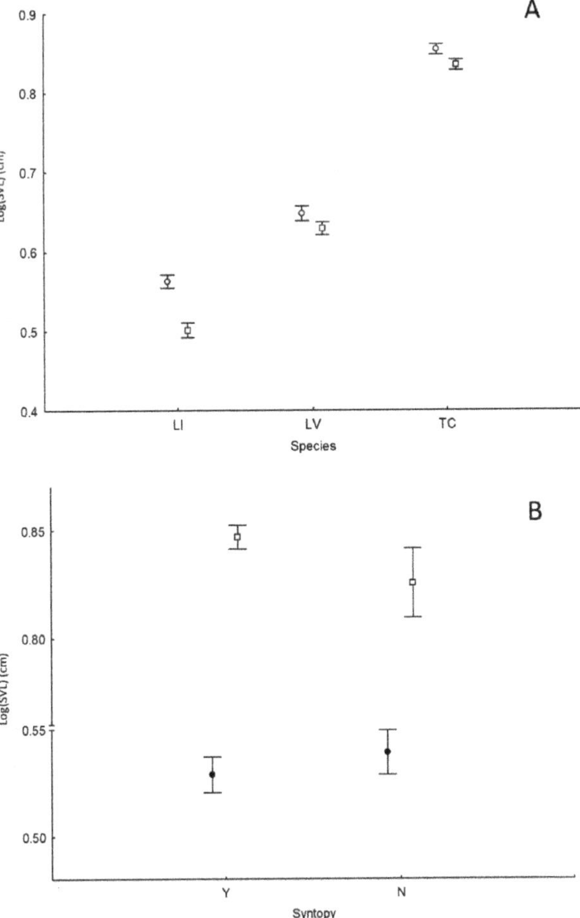

Figure 3. (**A**) Effect of species and sex on newt body size (SVL). LI = *L. italicus*; LV = *L. vulgaris*; TC = *T. carnifex*. Circles = females; squares = males. (**B**) Effect of syntopy on newt body size (SVL). Black circles = *L. italicus*; empty squares *T. carnifex*. Y = two or more species co-occurring; N = single species. *Lissotriton vulgaris* was excluded from this analysis because it was always found in syntopic condition with the remaining species. Vertical bars denote 0.95 confidence intervals.

3.2. Diet Analysis

Overall, few prey categories dominated the diet spectrum of the study species in terms of frequency (Table S3): insect larvae in *T. carnifex* (69%) and *L. italicus* (65%), and cladocerans in *L. vulgaris* (62%). *Lissotriton italicus* diet was dominated by aquatic insect larvae (32%) followed by cladocerans (23%). In *L. vulgaris*, the most important food types were cladocerans (72%). *Triturus carnifex* diet was dominated by cladocerans (48%). Overall, the average number of preys ingested by each newt was 21.53 ± 32.76 (mean ± SD) with minimum and maximum number per stomach ranging 3–383.

3.3. Individual Specialization and the Effect of Coexisting Species on Within-Population Diet Variation

In terms of Total Niche Width (TNW), newt populations did not show any interspecific difference ($F_{2,22}$ = 1.311, p = 0.290) or any variation in the different condition of syntopy ($F_{1,16}$ = 0.007, p = 0.935), with consistent behavior among species ($F_{1,16}$ = 1.724, p = 0.208). Within-population diet variation (V) significantly varied across sites, species (on average, *L. vulgaris* showed smaller V than the other species), and syntopy condition with V being significantly higher in coexisting populations (Table 2).

Table 2. Effect of site (random factor), species, sex, syntopy condition (yes/no), newt body size (SVL; covariate), and interaction terms of fixed factors on newt diet variation (V = 1 − IS; see Methods) at the study sites. The effect of syntopy condition (°) was tested on *L. italicus* and *T. carnifex* only because *L. vulgaris* was always found in co-occurrence with other species. Significant effects are highlighted in bold.

Effect	SS	DoF	MS	F	*p*
Site	**10.177**	17	**0.599**	**13.181**	**<0.001**
Species	**0.600**	2	**0.300**	**6.603**	**0.001**
Sex	0.038	1	0.038	0.835	0.361
Syntopy°	**0.677**	1	**0.677**	**10.305**	**0.001**
SVL	0.007	1	0.007	0.155	0.694
Species × Syntopy°	0.108	1	0.108	1.638	0.201
Species × Sex	0.049	2	0.024	0.534	0.586
Sex × Syntopy°	0.043	1	0.043	0.654	0.419
Error	22.165	488	0.045		

× Effects tested on *L. vulgaris* and *T. carnifex* only.

Moreover, V increased with population niche breadth (TNW) in all three newt species considered. Linear regression confirmed a significant positive slope in each case for both observed and simulated data, with observed V values always larger than simulated ones, whereas slopes did not differ between observed and simulated data (LI: $slope_{obs}$ = 0.305, R^2_{obs} = 0.743, $slope_{sim}$ = 0.194, R^2_{sim} = 0.656, obs ≠ sim $F_{1,12}$ = 1.426, p = 0.255. LV: $slope_{obs}$ = 0.269; R^2_{obs} = 0.882, $slope_{sim}$ = 0.317, R^2_{sim} = 0.938, obs ≠ sim $F_{1,6}$ = 0.429, p = 0.537; TC: $slope_{obs}$ = 0.243, R^2_{obs} = 0.461, $slope_{sim}$ = 0.248, R^2_{sim} = 0.702, obs ≠ sim $F_{1,20}$ = 0.003, p = 0.956). When we pooled data from all the species, the overall patterns of the relationship between V and TNW mirrored those found in each single species ($slope_{obs}$ = 0.273, R^2_{obs} = 0.611, $slope_{sim}$ = 0.239, R^2_{sim} = 0.742, obs ≠ sim $F_{1,48}$ = 0.401, p = 0.530. Figure 4A). This pattern was apparently disrupted by the effect of syntopy condition: populations occurring as a single species had a significantly stronger and steeper V vs. TNW relationship than the null model ($slope_{obs}$ = 0.329, R^2_{obs} = 0.930, $slope_{sim}$ = 0.231, R^2_{sim} = 0.795, obs ≠ sim $F_{1,12}$ = 4.848, p = 0.049. Figure 4C), whereas populations occurring syntopically with others exhibited a larger diet variation than the null model but with no significant relationship with TNW ($slope_{obs}$ = −0.014, R^2_{obs} = 0.002, $slope_{sim}$ = 0.242, R^2_{sim} = 0.700, obs ≠ sim $F_{1,30}$ = 0.001, p = 0.976. Figure 4B).

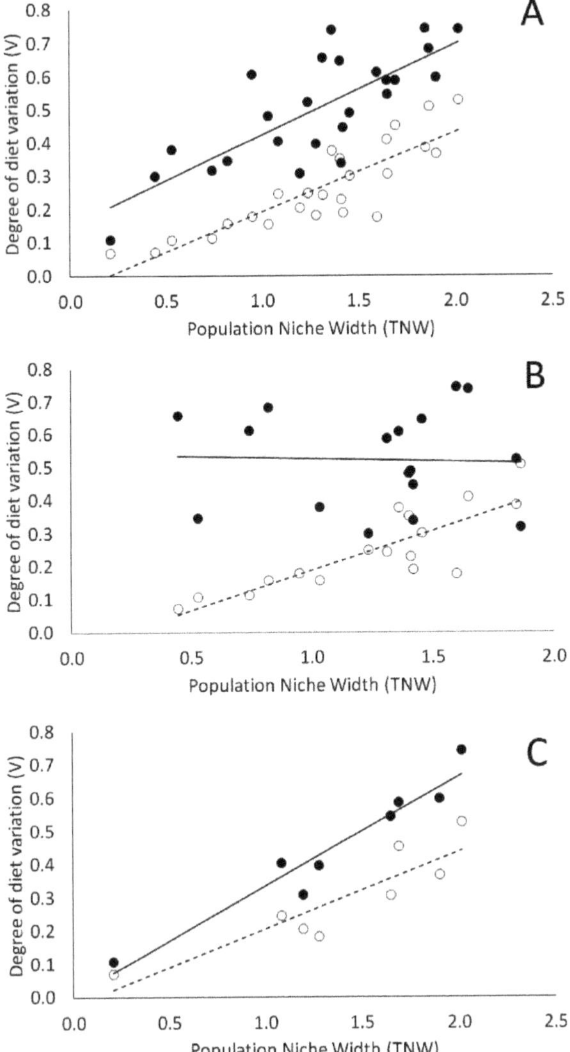

Figure 4. Correlation between diet variation among individuals (V) and the TNW for all the newt populations (**A**), for coexisting populations (**B**), and for population living in isolation (**C**). The observed values are shown with filled circles (and continue regression line). Empty circles (and the dotted regression line) indicate the expected trend under a null model in which diet variation arises solely by individuals randomly sampling a limited set of prey from a shared prey distribution.

As for the variation of the degree of individual specialization between species and syntopy condition, a clear consistent pattern emerged; the largest species (*T. carnifex*) showed a higher degree of individual specialization than the smallest one (*L. italicus*) and both species revealed a significantly higher percentage of specialized individuals in syntopic conditions (Table 3; Figure 5).

Table 3. Effect of species, syntopy, and interaction terms of fixed factors on the percentage of resampling PSi < observed Psi of *L. italicus* and *T. carnifex* populations (see Material and Methods). *Lissotriton vulgaris* was excluded from the analysis since it was always found in syntopic condition with other species. Significant effects are highlighted in bold.

Effect	SS	DoF	F	*p*
Intercept	0.360	1	83.692	0
Species	**0.022**	1	**5.176**	**0.037**
Syntopy	**0.0308**	1	**7.148**	**0.017**
Species × Syntopy	0.00003	1	0.008	0.930
Error	0.069	16		

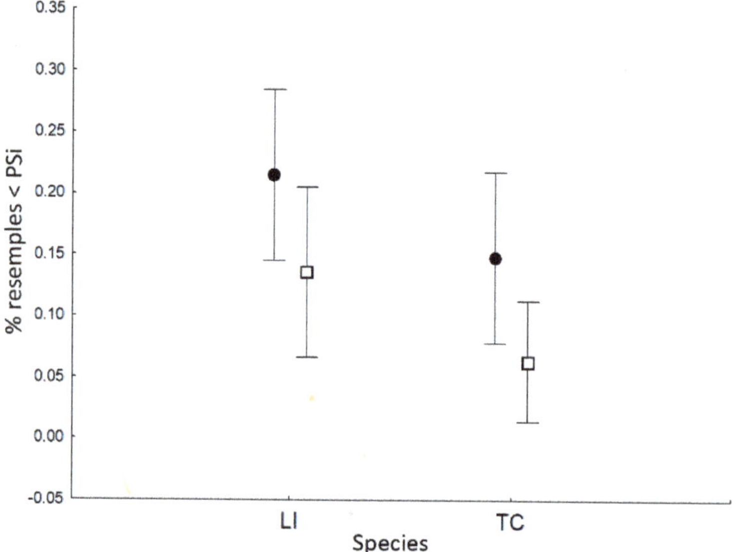

Figure 5. Effect of species and syntopy on the percentage of resampling PSi < observed Psi of the newt populations (see Methods). *Lissotriton vulgaris* was excluded from the analysis because this species was always found in syntopic condition with other newt species. Black circles = species alone; empty squares = two or more coexisting species; LI = *L. italicus*; TC = *T. carnifex*. Vertical bars denote 0.95 confidence intervals.

4. Discussion

4.1. Body Size and Diet

Newts are considered opportunistic and generalist predators [11,53]. Overall, our results confirm this suggestion as in our study all the three species exploited a large variety of resources with a clear prevalence of aquatic insects (larvae and pupae) and other aquatic invertebrates (mainly cladocerans).

The three studied newt species relied on the same prey types and their dietary similarity was high, as already shown in earlier studies [14,15,19,54]. Our findings are also consistent with previous studies documenting that syntopic species generally show overlap in the use of resources during the aquatic phase [19,54,55]. Newts also generally exhibit a seasonal dietary plasticity [21], which allows them to consume prey opportunistically in relation to their availability and local abundance [18,56].

As expected, the three species vary significantly in SVL, with *T. carnifex* being the largest and *L. italicus* the smallest [22]. Although this result is not surprising per se, it is, however, intriguing that *T. carnifex* grew larger when found in syntopy with one or two smaller newt species than elsewhere. It is possible that this enlarged body size may merely depend on a more abundant prey availability, allowing two or three species to live together. We reject this hypothesis; whereas prey availability should have affected all the species belonging to the same assemblage if the hypothesis was to be supported, we found that, instead, this pattern was not consistent across species (see below for *L. italicus*). Moreover, this pattern could also be due to an advantage of growing bigger as a larger body size may minimize niche overlap with the smaller species, by widening the trophic spectrum (i.e., preying on bigger prey items; [33]), or even to increase the ability to directly prey on the smaller ones (we found a subadult of *L. vulgaris* in a *T. carnifex* stomach; [19,57]). Our data are not sufficient to disentangle the true reasons behind the above-mentioned pattern. Italian newts were larger in absence of the other two species. Since the three wells where we found *L. italicus* as single species were very small and similar between each other, we can exclude the absolute size of the site as the factor. It is possible that interspecific interactions, such as competition for trophic resources, can impact on the body size of *L. italicus*, on the other hand smaller sizes may allow the usage of resources not reachable by the larger species (e.g., refugees to escape predation, etc.).

4.2. Individual Specialization and the Effect of Coexisting Species on Within-Population Diet Variation

In our study system, we assumed that a newt population living in absence of other newt species can access resources that may otherwise be depleted or monopolized by competitors, which thus can experience a niche expansion via "ecological release" [27]. This niche width (NW) variation may occur at both the population and the individual levels. Bolnick et al. [27] illustrated three scenarios with (1) a population NW varying together with individual NWs (parallel release), (2) a population NW varying while individual NWs remain constant (NVH; [58]), (3) a population NW remaining constant and individual NWs expanding ("individual release"). In our study system, newt populations did not show interspecific differences in their population NW, which in turn did not vary between presence and absence of coexisting species (i.e., competitors). Overall, the within-population diet variation (V) increased with population NW in all three newt species examined, showing values of V significantly larger than null populations but with no significant slopes. This would mean that the observed pattern could be explained by the null model alone. The increasing values of V at larger values of TNW was potentially due to the small number of preys recorded per individual, which may have underestimated the diversity of preys actually consumed, and therefore, overestimated the variation among individuals [27]. However, it is unlikely that the high number of prey per stomach recorded in all the three species did not properly represent the actual diversity of prey consumed by newts. In addition, the observed level of diet variation, always significantly larger than expected by chance, indicated a high degree of individual specialization not explained by the null model (random sampling from a common diet distribution) [26].

The observed positive relationship between the degree of diet variation and the population diet width was consistent among all the populations considered, but this pattern was disrupted by the condition of syntopy. In presence of coexisting species, newt populations showed an individual diet variation invariant with TNW. This would suggest that interspecific competition could limit the degree of individual diet variation by setting a maximum threshold. In this condition, the increase of TNW is accomplished by the increase of individual NW. Conversely, when newt species did not face potential interspecific competitors, individuals showed larger diet variations, with the relationship with TNW being significantly steeper than the null model. *L. italicus* and *T. carnifex* showed a consistent pattern of higher individual specialization in absence of coexisting heterospecific newts, and therefore, a comparatively lower individual NW. Theoretical models predict that niche variation between individuals should weaken coexistence by reducing (a) species-level niche differentiation and (if coupled with demographic stochasticity) (b) the likelihood of long-term coexistence by favoring abundant competitors over species recovering from small population sizes [29]. Our short-term study does not exactly mirror these predictions but revealed that interspecific competition could limit the degree of trophic niche variation between individuals and thus favoring species coexistence. By and large, the observed pattern of newt behavior (stable TNW and increased individual specialization) was not consistent with any of the proposed scenarios.

Costa-Pereira et al. [30], in a field study on coexisting frogs in central Brazil, suggested that individual niche specialization can be strongly context-dependent within and across species. Thus, the hierarchies of individual variation among coexisting species are not necessarily consistent across communities. Similarly, Cloyed and Eason [31] found inconsistent patterns for the effect of heterospecific density on individual trophic specialization. Our study was partially in disagreement with this conclusion, as in our newt assemblages there was a consistent pattern with the degree of individual specialization being (1) size-dependent (percentage of specialized individuals increasing within larger sized species) and (2) assemblage-complexity-dependent (percentage of specialized individuals increasing when a given species does occur syntopically with other species in comparison when it occurs singly in the environment). Therefore, our study provided more consistent evidence with the experimental study on freshwater fish by Bolnick et al. [27] in showing a heterogeneous and contrasting effect of competition on the individual specialization and on the niche width of the whole population.

5. Conclusions

Population niche breadth is thought to represent a balance between the diversifying effect of intraspecific competition and the constraints imposed by interspecific competition [51]. Our study system, consisting of several replicates of the same habitat hosting different co-occurring species combinations, represents a unique opportunity to study the causes and consequences of population and individual niche variation in natural communities. Indeed, our study elucidates important aspects of how individual niche specialization varies across similar coexisting species. While our findings support some predictions of current theory (i.e., the diffuse occurrence of individual specialization [32]), they also provide novel insights into how individual niche variation in closely related and ecologically similar species diverges in different conditions of syntopy. Moreover, while most studies concluded that the ecological parameters driving the development of individual specialization are not consistent across potentially competing species [30,31], we found consistent pattern of increased individual specialization across species in presence of interspecific competition.

Since all the studies published so far consisted of only few sympatric species (up to three in our study case, up to four in Costa-Pereira et al. [30], up to five in Cloyed and Eason [31], and even just two in the experiments by Bolnick et al. [27]), it remains unstudied how much the "intensity" and the "regime" of the interspecific specialization may vary in natural species-rich communities and at different levels of the trophic chain. Without more studies on species-rich communities, it will remain impossible to effectively understand whether the full NVH may be accepted or rejected. In addition, in our study we did not analyze newt density and prey availability as factors potentially influencing the observed patterns. However, since some of the patterns described in this paper may be density dependent [30] or affected by ecological opportunity [59], further studies are needed to examine the effects of population densities and prey availability on population and individual niche variation.

Supplementary Materials: The following are available online at http://www.mdpi.com/1424-2818/12/5/181/s1, Table S1: List of the surveyed sites with the recorded species and geographic coordinates. Table S2: Synopsis of the all sampled population in all surveyed sites in the study area. Species codes: Li = *Lissotriton italicus*; Lv = *Lissotriton vulgaris*; Tc = *Triturus carnifex*. Sex codes: M = males; F = females; N = number of sampled individuals; SVL = average snout-vent-length (cm). Table S3: Number (N) and frequency of occurrence (Freq) of prey items ingested by the three species studied. Species codes: Li = *Lissotriton italicus*; Lv = *Lissotriton vulgaris*; Tc = *Triturus carnifex*. The number of analyzed stomachs per species is reported in brackets.

Author Contributions: Conceptualization, L.V.; methodology, L.V.; software, L.V. and J.M.; validation, A.M.B. and L.L.; formal analysis, L.V. and J.M.; investigation, J.M. and L.S.; resources, L.S. and L.V.; data curation, J.M. and A.M.B.; writing—original draft preparation, A.M.B., L.L., and L.V.; writing—review and editing, M.A.B., L.S., and J.M.; visualization, A.M.B.; supervision, L.V.; All authors have read and agreed to the published version of the manuscript.

Funding: This research was partly funded by the Grant to Department of Science, Roma Tre University (MIUR-Italy Dipartimenti di Eccellenza, ARTICOLO 1, COMMI 314–337 LEGGE 232/2016).

Acknowledgments: The authors are indebted to the Aurunci Natural Park, which authorized this research, with special thanks to G. De Marchis and C. Esposito for their help. We also thank G. Simbula for her help in the field.

Conflicts of Interest: The authors declare no conflict of interest.

References

1. Petranka, J.W. Fish predation: A factor affecting the spatial distribution of a stream-breeding salamander. *Copeia* **1983**, *1983*, 624–628. [CrossRef]
2. Resetarits, W.J. Differences in an ensemble of streamside salamanders (Plethodontidae) above and below a barrier to brook trout. *Amphibia-Reptilia* **1997**, *18*, 15–25. [CrossRef]
3. Wilkins, R.N.; Peterson, N.P. Factors related to amphibian occurrence and abundance in headwater streams draining second-growth Douglas-fir forests in southwestern Washington. *For. Ecol. Manag.* **2000**, *139*, 79–91. [CrossRef]
4. Sánchez-Hernández, J. Reciprocal role of salamanders in aquatic energy flow pathways. *Diversity* **2020**, *12*, 32. [CrossRef]
5. Murphy, M.L.; Hall, J.D. Vaired effects of clear-cut logging on predators and their habitat in small streams of the Cascade Mountains, Oregon. *Can. J. Fish. Aq. Sci.* **1981**, *38*, 137–145. [CrossRef]
6. Marcot, B.G.; Vander Heyden, M. Key ecological functions of wildlife species. In *Wildlife-Habitat Relationships in Oregon and Washington*; Johnson, D.H., O'Neil, T.A., Eds.; Technical Coordinators Oregon State University Press: Corvallis, OR, USA, 2001.
7. Schabetsberger, R.; Jersabek, C.D. Alpine newts (*Triturus alpestris*) as top predators in a high-altitude karst lake: Daily food consumption and impact on the copepod *Arctodiaptomus Alpinus*. *Freshw. Biol.* **1995**, *33*, 47–61. [CrossRef]
8. Cogălniceanu, D.; Palmer, M.; Ciubuc, C. Feeding in anuran communities on islands in the Danube floodplain. *Amphibia-Reptilia* **2001**, *22*, 1–19.
9. Buono, V.; Bissattini, A.M.; Vignoli, L. Can a cow save a newt? The role of cattle drinking troughs in amphibian conservation. *Aquat. Conserv.* **2019**, *29*, 964–975. [CrossRef]
10. Semlitsch, R.D.; O'Donnell, K.M.; Thompson, F.R., III. Abundance, biomass production, nutrient content, and the possible role of terrestrial salamanders in Missouri Ozark forest ecosystems. *Can. J. Zool.* **2014**, *92*, 997–1004. [CrossRef]
11. Wells, K.D. *The Ecology and Behavior of Amphibians*; University of Chicago Press: Chicago, IL, USA, 2007.
12. Davic, R.D.; Welsh JR, H.H. On the ecological roles of salamanders. *Annu. Rev. Ecol. Evol. Syst.* **2004**, *35*, 405–434. [CrossRef]
13. Polis, G.A.; Strong, D.R. Food web complexity and community dynamics. *Am. Nat.* **1996**, *147*, 813–846. [CrossRef]
14. Avery, R.A. Food and feeding relations of three species of *Triturus* (Amphibia Urodela) during the aquatic phases. *Oikos* **1968**, *19*, 408–412. [CrossRef]
15. Griffiths, R.A. Feeding niche overlap and food selection in smooth and palmate newts, *Triturus vulgaris* and *T. helveticus*, at a pond in mid-Wales. *J. Anim. Ecol.* **1986**, *55*, 201–214. [CrossRef]
16. Joly, P.; Giacoma, C. Limitation of similarity and feeding habits in three syntopic species of newts (*Triturus*, Amphibia). *Ecography* **1992**, *15*, 401–411. [CrossRef]
17. Petranka, J.W. *Salamanders of the United States and Canada*; Smithsonian Institution Press: Washington, DC, USA, 1998.
18. Vignoli, L.; Bologna, M.A.; Luiselli, L. Seasonal patterns of activity and community structure in an amphibian assemblage at a pond network with variable hydrology. *Acta Oecol.* **2007**, *31*, 185–192. [CrossRef]
19. Vignoli, L.; Luiselli, L.; Bologna, M.A. Dietary patterns and overlap in an amphibian assemblage at a pond in Mediterranean Central Italy. *Vie Milieu* **2009**, *59*, 47–57.
20. Kopecký, O.; Novak, K.; Vojar, J.; Šusta, F. Food composition of alpine newt (*Ichthyosaura alpestris*) in the post-hibernation terrestrial life stage. *North-West. J. Zool.* **2016**, *12*, 299–303.
21. Fasola, M.; Canova, L. Feeding habits of *Triturus vulgaris*, *T. cristatus* and *T. alpestris* (Amphibia, Urodela) in the northern Apennines (Italy). *Ital. J. Zool.* **1992**, *59*, 273–280.
22. Lanza, B.; Andreone, F.; Bologna, M.A.; Corti, C.; Razzetti, E. *Amphibia*; Edizioni Calderini: Bologna, Italy, 2007.
23. Bologna, M.A.; Capula, M.; Carpaneto, G.M. *Anfibi e Rettili Del Lazio*; Palombi Editori: Rome, Italy, 2000.

24. Araújo, M.S.; Bolnick, D.I.; Layman, C.A. The ecological causes of individual specialisation. *Ecol. Lett.* **2011**, *14*, 948–958. [CrossRef]

25. Bolnick, D.I.; Ballare, K.M. Resource diversity promotes among-individual diet variation, but not genomic diversity, in lake stickleback. *Ecol. Lett.* **2020**, *23*, 495–505. [CrossRef]

26. Bolnick, D.I.; Svanbäck, R.; Araújo, M.S.; Persson, L. Comparative support for the niche variation hypothesis that more generalized populations also are more heterogeneous. *Proc. Natl. Acad. Sci. USA* **2007**, *104*, 10075–10079. [CrossRef] [PubMed]

27. Bolnick, D.I.; Ingram, T.; Stutz, W.E.; Snowberg, L.K.; Lau, O.L.; Paull, J.S. Ecological release from interspecific competition leads to decoupled changes in population and individual niche width. *Proc. R. Soc. B* **2010**, *277*, 1789–1797. [CrossRef] [PubMed]

28. Barabás, G.; D'Andrea, R. The effect of intraspecific variation and heritability on community pattern and robustness. *Ecol. Lett.* **2016**, *19*, 977–986. [CrossRef] [PubMed]

29. Hart, S.P.; Schreiber, S.J.; Levine, J.M. How variation between individuals affects species coexistence. *Ecol. Lett.* **2016**, *19*, 825–838. [CrossRef] [PubMed]

30. Costa-Pereira, R.; Rudolf, V.H.; Souza, F.L.; Araújo, M.S. Drivers of individual niche variation in coexisting species. *J. Anim. Ecol.* **2018**, *87*, 1452–1464. [CrossRef] [PubMed]

31. Cloyed, C.S.; Eason, P.K. Different ecological conditions support individual specialization in closely related, ecologically similar species. *Evol. Ecol.* **2016**, *30*, 379–400. [CrossRef]

32. Bolnick, D.I.; Svanbäck, R.; Fordyce, J.A.; Yang, L.H.; Davis, J.M.; Hulsey, C.D.; Forister, M.L. The ecology of individuals: Incidence and implications of individual specialization. *Am. Nat.* **2003**, *161*, 1–28. [CrossRef]

33. Vignoli, L.; Bissattini, A.M.; Luiselli, L. Food partitioning and the evolution of non-randomly structured communities in tailed amphibians: A worldwide systematic review. *Biol. J. Lin. Soc.* **2017**, *120*, 489–502. [CrossRef]

34. Cloyed, C.S.; Eason, P.K. Feeding limitations in temperate anurans and the niche variation hypothesis. *Amphibia-Reptilia* **2017**, *38*, 473–482. [CrossRef]

35. Toscano, B.J.; Gownaris, N.J.; Heerhartz, S.M.; Monaco, C.J. Personality, foraging behavior and specialization: Integrating behavioral and food web ecology at the individual level. *Oecologia* **2016**, *182*, 55–69. [CrossRef]

36. Cerini, F.; Bologna, M.A.; Vignoli, L. Dragonflies community assembly in artificial habitats: Glimpses from field and manipulative experiments. *PLoS ONE* **2019**, *14*. [CrossRef] [PubMed]

37. Scillitani, G.; Scalera, R.; Carafa, M.; Tripepi, S. Conservation and biology of *Triturus italicus* in Italy (Amphibia, Salamandridae). *It. J. Zool.* **2004**, *71*, 45–54. [CrossRef]

38. Razzetti, E.; Bernini, F. Triturus vulgaris. In *Atlas of Italian Amphibians and Reptiles Societas Herpetologica Italica*; Sindaco, R., Doria, G., Razzetti, E., Bernini, F., Eds.; Edizioni Polistampa: Firenze, Italy, 2006.

39. Novellino, D. An account of basket weaving and the use of fibre plants in the Mount Aurunci Regional Park (Central Italy). In Proceedings of the Fourth International Conference of Ethnobotany (ICEB 2005), Istanbul, Turkey, 21–26 August 2005.

40. Anzalone, B. *Prodromo Della Flora Romana: Elenco Preliminare Delle Piante Vascolari Spontanee Del Lazio*; Regione Lazio: Rome, Italy, 1984.

41. Minutillo, F.; Moraldo, B.; Ross, W. Segnalazioni floristiche italiane. *Inform. Bot. Ital.* **1985**, *17*, 124–127.

42. Romano, A.; Montinaro, G.; Mattoccia, M.; Sbordoni, V. Amphibians of the Aurunci Mountains (Latium, Central Italy). Checklist and conservation guidelines. *Acta Herpetol.* **2007**, *2*, 17–25.

43. Salvidio, S.; Sindaco, R.; Emanueli, L. Feeding habits of sympatric *Discoglossus montalentii, Discoglossus sardus* and *Euproctus montanus* during the breeding season. *Herpetol. J.* **1999**, *9*, 163–167.

44. Lizana, M.; Perez-Mellado, V.; Ciudad, M.J. Analysis of the structure of an amphibian community in the central system of Spain. *Herpetol. J.* **1990**, *1*, 435–446.

45. Reques, R.; Tejedo, M. Fenología y hábitats reproductivos de una comunidad de anfibios en la Sierra de Cabra (Córdoba). *Rev. Esp. Herp.* **1991**, *6*, 49–54.

46. Solé, M.; Beckmann, O.; Pelz, B.; Kwet, A.; Engels, W. Stomach-flushing for diet analysis in anurans: An improved protocol evaluated in a case study in Araucaria forests, southern Brazil. *Stud. Neotrop. Fauna Environ.* **2005**, *40*, 23–28. [CrossRef]

47. Newsome, S.D.; Tinker, M.T.; Gill, V.A.; Hoyt, Z.N.; Doroff, A.; Nichol, L.; Bodkin, J.L. The interaction of intraspecific competition and habitat on individual diet specialization: A near range-wide examination of sea otters. *Oecologia* **2015**, *178*, 45–59. [CrossRef]

48. Bolnick, D.I.; Yang, L.H.; Fordyce, J.A.; Davis, J.M.; Svanbäck, R. Measuring individual-level resource specialization. *Ecology* **2002**, *83*, 2936–2941. [CrossRef]

49. Schoener, T.W. The Anolis lizards of Bimini: Resource partitioning in a complex fauna. *Ecology* **1968**, *49*, 704–726. [CrossRef]

50. Svanbäck, R.; Persson, L. Individual diet specialization, niche width and population dynamics: Implications for trophic polymorphisms. *J. Anim. Ecol.* **2004**, *73*, 973–982. [CrossRef]

51. Roughgarden, J. Evolution of niche width. *Am. Nat.* **1972**, *106*, 683–718. [CrossRef]

52. Feinsinger, P.; Spears, E.E.; Poole, R.W. A simple measure of niche breadth. *Ecology* **1981**, *62*, 27–32. [CrossRef]

53. Griffiths, R.A. *Newts and Salamanders of Europe*; T & AD Poyser: London, UK, 1996.

54. Dolmen, D.; Koksvik, J.I. Food and feeding habits of *Triturus vulgaris* (L.) and *T. cristatus* (Laurenti) (Amphibia) in two bog tarns in central Norway. *Amphibia-Reptilia* **1983**, *4*, 17–24. [CrossRef]

55. Griffiths, R.A.; Mylotte, V.J. Microhabitat selection and feeding relations of smooth and warty newts, *Triturus vulgaris* and *T. cristatus*, at an upland pond in mid-Wales. *Ecography* **1987**, *10*, 1–7. [CrossRef]

56. Vignoli, L.; Bombi, P.M.; D'Amen, M.; Bologna, M.A. Seasonal variation in the trophic niche of a heterochronic population of *Triturus alpestris apuanus* from the south-western Alps. *Herpetol. J.* **2007**, *17*, 183–191.

57. Fasola, M. Resource partitioning by three species of newts during their aquatic phase. *Ecography* **1993**, *16*, 73–81. [CrossRef]

58. Van Valen, L. Morphological variation and width of ecological niche. *Am. Nat.* **1965**, *99*, 377–390. [CrossRef]

59. Salvidio, S.; Crovetto, F.; Costa, A. Individual trophic specialisation in the Alpine newt increases with increasing resource diversity. *Ann. Zool. Fenn.* **2019**, *56*, 17–24. [CrossRef]

 *diversity*

Article

Same Diet, Different Strategies: Variability of Individual Feeding Habits across Three Populations of Ambrosi's Cave Salamander (*Hydromantes ambrosii*)

Enrico Lunghi [1,2,*], Fabio Cianferoni [2,3], Filippo Ceccolini [2], Yahui Zhao [1,*], Raoul Manenti [4], Claudia Corti [2], Gentile Francesco Ficetola [4,5] and Giorgio Mancinelli [6,7,8]

1   Key Laboratory of the Zoological Systematics and Evolution, Institute of Zoology, Chinese Academy of Sciences, Beijing 100101, China
2   Museo di Storia Naturale dell'Università degli Studi di Firenze, Museo "La Specola", 50125 Firenze, Italy; cianferoni.fabio@gmail.com (F.C.); ceccolinif@virgilio.it (F.C.); claudia.corti@unifi.it (C.C.)
3   Istituto di Ricerca sugli Ecosistemi Terrestri, Consiglio Nazionale delle Ricerche, 50010 Sesto Fiorentino (Firenze), Italy
4   Dipartimento di Scienze e Politiche Ambientali, Università degli Studi di Milano, 20133 Milano, Italy; raoulmanenti@gmail.com (R.M.); francesco.ficetola@gmail.com (G.F.F.)
5   Laboratoire d'Écologie Alpine (LECA), CNRS, University Grenoble Alpes, 38000 Grenoble, France
6   Dipartimento di Scienze e Tecnologie Biologiche ed Ambientali (DiSTeBA), Università del Salento, 73100 Lecce, Italy; giorgio.mancinelli@unisalento.it
7   Istituto per le Risorse Biologiche e le Biotecnologie Marine (IRBIM), Consiglio Nazionale delle Ricerche (CNR), 08040 Lesina (Foggia), Italy
8   CoNISMa, Consorzio Nazionale Interuniversitario per le Scienze del Mare, 00196 Roma, Italy
*   Correspondence: enrico.arti@gmail.com (E.L.); zhaoyh@ioz.ac.cn (Y.Z.)

Received: 14 March 2020; Accepted: 3 May 2020; Published: 6 May 2020

Abstract: European cave salamanders of the genus *Hydromantes* are a group of eight species endemic to Italy and south-eastern France. Knowledge on the trophic niche of European *Hydromantes* is poor, and the few available studies only partially investigate their feeding habits. We performed an in-depth study on the trophic niche of the Ambrosi's cave salamander (*H. ambrosii*), assessing the potential divergences among three different populations. All the populations had a similar diet composition, showing a wider trophic niche in fall compared to spring. In only one population, "true specialists" were present; however, in all three populations, generalist individuals always represented the larger proportion. Interspecific and intraspecific competition did not play an important role in determining individual dietary specialisation in *H. ambrosii*; contrarily, the characteristics of the surrounding environment seemed to be an important factor. The best body conditions were observed in the population located in the site where the non-arboreal vegetation cover was the highest. Besides providing new information on the trophic niche of *H. ambrosii*, we here showed that studies encompassing both intrinsic and extrinsic factors at the population level are needed to fully understand the trophic dynamics occurring among European cave salamanders.

Keywords: individual diet specialization; ecological opportunity; diet; plethodontid; cave biology

1. Introduction

European plethodontid salamanders (genus *Hydromantes*; see [1] for taxonomic discussion) are composed of eight species distributed in Italy and in Provencal France [2]. European *Hydromantes* all show allopatric distributions [2,3] with one exception. Two mainland species, *H. ambrosii* and *H. italicus*, overlap in a small area corresponding, respectively, to their most southern and northern distributions;

this has naturally led to the occurrence of hybrid populations [4]. Although not being strictly cave species [2], the se relatively small salamanders are often found in caves and other subterranean habitats, where they seek specific microclimatic conditions [5]. Plethodontid salamanders lack lungs and mainly breath through the skin [6]; to keep their respiration efficient, the y need relatively low temperatures and high moisture [7]. This particular microclimate is found in subterranean habitats all year round [8,9]; accordingly, stable populations of *Hydromantes* are often found in these habitats, where they can reach high abundances [2]. Additionally, besides the high environmental suitability, subterranean habitats generally lack predators ([10]; but see [11,12] for a few exceptions), a condition allowing *Hydromantes* salamanders to select these habitats for their long reproductive period [13].

Feeding ecology is an ecological trait of these salamanders that has recently received special attention. The absence of lungs in *Hydromantes* enabled the evolution of a protrusible tongue that is used to catch a wide range of prey [6,14,15]. Indeed, *Hydromantes* mostly adopt a sit-and-wait hunting strategy, waiting until suitable prey come within reach of the tongue, which can be extended up to 80% of the salamander's body length [16,17]. Prey diversity in *Hydromantes* is high, ranging from flying insects to aquatic larvae, comprising at least 35 different invertebrate orders [15,18,19]. Moreover, the se salamanders often "recycle" their own organic matter such as unfertilized eggs or the skin after moulting [15]. Foraging activity is more intense during spring and fall [15,19,20], seasons in which the appropriate climatic conditions allow to leave the subterranean refuges and exploit outdoor environments, where prey availability is higher [21,22]. Following prey capture, salamanders return to their refuges where suitable microclimatic conditions ensure effective cutaneous respiration but also promote digestion [7,23,24]. A comparative study including six European *Hydromantes* species (*H. ambrosii*, *H. flavus*, *H. supramontis*, *H. imperialis*, *H. sarrabusensis* and *H. genei*) indicated that, although few prey categories account for the highest proportion of food items in all of them, a difference in diet composition between these species occurs [25]. No differences in diet composition and prey diversity occurred between adults and juveniles, but the latter usually consumed fewer and smaller prey, a constraint likely due to their size [25,26]. However, the authors did not assess any potential intraspecific divergence in diet composition.

Individual diet specialization (IS) occurs when individuals use a subset of prey types included in the population's trophic niche [26,27]. As a consequence, the niche width of the population is determined by both the intra-individual dietary diversity, and by the variability of prey types in the diet of the different individuals [28,29]. IS has major impacts on population stability, interspecific interactions and food web structure [30,31]; nevertheless the occurrence of IS has been evaluated only for a limited number of species (see [32–34] for some examples). The presence of individual specialisation in the use of trophic resources has been assessed in most of the European *Hydromantes*. Within the six studied species (*H. strinatii*, *H. flavus*, *H. supramontis*, *H. imperialis*, *H. sarrabusensis* and *H. genei*) each population had "true specialists", although in different proportions [35–37]. Individuals may specialise to use a subset of the available resources as a consequence of negative biological interactions, because a high prey diversity promotes individuals' preference, or of both conditions acting synergistically [26,38]. The refore, the full spectrum of resources consumed by a population (the total niche width) can be explained by the variability of the resources consumed by each individual (within-individual component), but it also includes the diversity occurring between individuals (between-individual component), as they can significantly differ from each other in the use of the available resources [27,39,40]. Individual diet specialization refers to the trophic strategy adopted by the single individuals regardless of age, sex or any other morphological constraint [31,33]. The studies conducted on *Hydromantes* indicated that the overall generalism of the species was due to a higher contribution of specialised individuals [35–37]. Beside the important preliminary information provided, some of these studies were limited in time (i.e., a single season [36]) and space (i.e., a single population [35]), a condition limiting the knowledge on the potential variability occurring among conspecific populations [37].

The aim of the present study was to perform an in-depth analysis of the trophic habits of the Ambrosi's cave salamander, *H. ambrosii*, which is one of the two *Hydromantes* species in which the individual diet specialization has never been explored before. Although the analysis is performed on one species, the re are very few studies investigating the individual variability of the feeding habits in multiple populations of the same species [37,41,42]), leaving a knowledge gap related to the potential variability occurring among conspecific populations. The stomach content of individuals from three Ligurian populations (NW Italy), collected in spring and fall, was analysed to test whether the dietary habits varied spatially and seasonally, and whether sexual, ontogenetic or size-related factors are playing a role. Two different methodological approaches were adopted to analyse these data: the first was based on conventional multivariate procedures, while the second relied on the estimation of niche metrics and two indices of individual specialization [27]. Specifically, we verified whether different populations showed comparable trophic niches and whether they were characterised by a similar proportion of specialised individuals. The analyses were repeated in spring and fall to assess whether the characteristics of populations' trophic niches change according to the season. In addition, we considered a diverse set of population features (i.e., body condition and density) as well as the environmental characteristics of the foraging area of the populations (i.e., vegetation cover of the cave surroundings) to evaluate their potential effects on the salamanders' trophic niches.

2. Methods

2.1. Study Species and Data

Hydromantes ambrosii is one of the eight species of European plethodontids, and its range occurs between Liguria and Tuscany, in north-western Italy [2]. The data analysed here were published by Lunghi et al. (2018) [15], reporting the stomach contents of a total of 124 individuals of *H. ambrosii* (67 females, 42 males and 15 juveniles) obtained from three different caves located in the La Spezia Province (Liguria, Italy; Figure 1) in 2016 and 2017; salamanders with empty stomach (38) and with only one prey item (31) were not considered in the analysis. The caves were far from the hybrid zone between *H. ambrosii* and the Italian cave salamander, *H. italicus* [4]; therefore, no congeneric (or hybrid) competitors occurred in the study area. Two of the surveyed sites (Populations 2 and 3) were relatively close (linear distance of about 180 m; Figure 1). Despite these sites being not far from each other, the available data suggest that the dispersal of *Hydromantes* mostly occurs over shorter distances (<100 m) [43,44]. Maximum dispersal distances <100 m are frequent in plethodontid salamanders [45]; consequently, we assumed the three populations to be independent.

Three samplings were performed, two in spring (2016 and 2017) and one in fall (2016). The sample size was relatively small and, to be able to employ all three populations in our analysis, we assumed that no seasonal niche breath variation occurred in the populations between different years (spring 2016 vs. 2017), thus we merged spring data to analyse at least 9 individuals for each population. Multiple samplings were performed to increase the robustness of the collected data [37,47]. Populations were haphazardly sampled as salamanders cannot be individually recognised. The study populations are very large (about 80 individuals for Population 3 and hundreds for Populations 1 and 2; [48]) and, under these conditions, the recapture rate of *Hydromantes* is, on average, ~0.5 [48]; therefore, repeated samplings of the same individuals were unlikely. A high rate of pseudo-replication would have happened if populations were repeatedly sampled during the same period, with a consequent increase in biased results. We are confident that the single seasonal survey performed here avoided this problem. Captured salamanders were stomach flushed and measured (snout-vent length (SVL), in mm) [15]; all salamanders showing an SVL < 40 mm were considered juveniles, while for adults, the sex was assessed basing on the presence/absence of male secondary sexual characteristics [2].

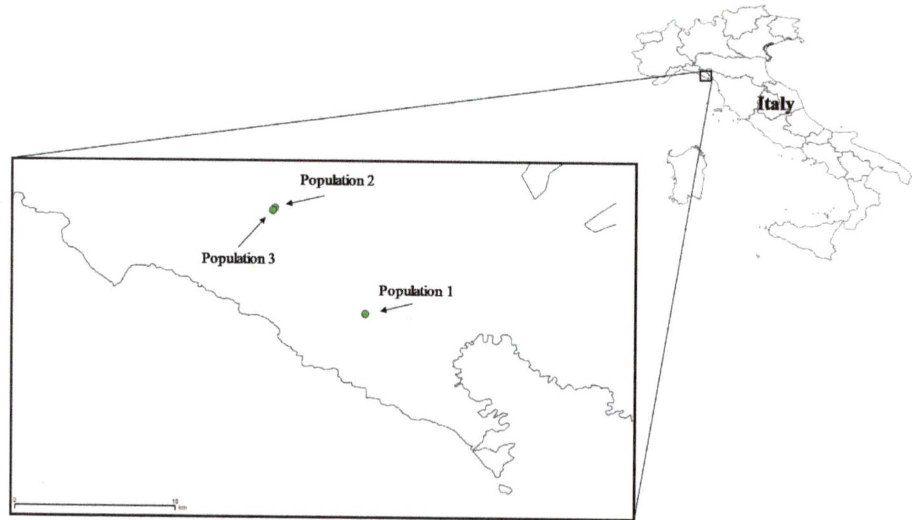

Figure 1. Location of the three studied populations of *Hydromantes ambrosii*. Green circles indicate the monitored caves. Further details are not shown for species protection [46].

Prey items were generally identified at the order level, and when possible, different life stages or specific families were also considered as independent groups, obtaining data related to 381 prey items belonging to 30 different prey groups [15]. All vertebrate food items were pooled into a single prey type (*Hydromantes*), and the prey categories were reduced to 28. Stomach contents always contained a certain amount of unidentified items for which a further prey category has been created. Only salamanders with at least two different prey categories in their stomachs were included in the analyses, to reduce the overestimation of individual specialization indices (see further in this section; [27]).

During the last survey (spring 2017), prior to undergoing stomach flushing, 74 randomly captured salamanders (24.67 ± 12.66 salamanders per cave; females = 37, males = 23, juveniles = 14) were measured in their total length (TL) and weighed (electronic scale, 0.01 g) to estimate their body condition index (hereafter, BCI). Only non-gravid females were used to estimate the BCI [13]. As the BCI, we used the Residual index [49,50]; residuals (i.e., the difference between the observed and the expected body mass) were estimated according to a log-log relationship between salamanders' weights and sizes. In this analysis we used the total length as these salamanders usually store fat in the tail [51].

Estimates of the abundance of these populations were obtained using N-mixture models, on the basis of repeated counts performed in spring 2017 [48]. For details on sampling procedures, data analyses and the reliability of the abundance estimates see reference [48].

We used these estimations to assess the density of the three populations (as abundance/surveyed area in m^2). In addition, the coordinates of the three *H. ambrosii* populations were used to extract quantitative information on the arboreal and non-arboreal vegetation cover (expressed in %) of the cave surroundings (area with diameter pixel size of 250 m) from the Terra MODIS Vegetation Continuous Field (VCF) product (available as MOD44B v006 at https://lpdaac.usgs.gov/products/mod44bv006/) for the years 2016 and 2017. Despite being not far from each other, Populations 2 and 3 are in an area with strong topographic and habitat heterogeneity and fall in different pixels of the MODIS vegetation maps. Arboreal vegetation cover included all forest types and age classes, while non-arboreal vegetation cover included meadows, regeneration areas and clear-cut areas.

2.2. Statistical Analysis

We used non-metric multi-dimensional (nMDS) analysis to explore the dietary similarity among the three populations under analysis. A two-way permutational multivariate analysis of variance (PERMANOVA with 10,000 permutations) was subsequently performed to test for the effects of the factor "season" (fixed, two levels, "spring" and "fall") and "ontogenetic stage" (fixed, two levels, "adult" and "juvenile") with the SVL of individual salamanders as a continuous covariate. PERMANOVA tests were repeated for adults only, verifying the influence of sex.

A population's total niche width (TNW) was calculated using the measure proposed by Roughgarden [52], based on the Shannon–Weaver diversity index:

$$\text{TNW} = -\sum_j q_j ln q_j \tag{1}$$

where q_j is the frequency of prey category j in the population's niche. TNW equals zero when all the individuals in the population consume only one single prey category and increases with both the number of prey categories and the evenness with which they are consumed. For each population, together with TNW, two additional niche metrics were estimated, i.e., the within- and between-individual components (WIC and BIC, respectively; [28,29]). In brief, the total niche width (TNW) of a population is assumed to be determined by the interaction of intra-individual variations in resource use (WIC), measured by the average variance of individuals' utilization functions $\rho(x|y)$ along a continuous dimension x

$$\text{WIC} = \overline{VAR\left[\rho(x|y)\right]} \tag{2}$$

and of inter-individual variations in resource use (BIC), measured by the variance of the frequency distribution for different individual averages in the population:

$$\text{BIC} = \overline{VAR[p(y)]} \tag{3}$$

where $p(y)$ is the frequency distribution of individuals with average use y in the population. Accordingly, the total niche width TNW represents the variance of the population's resource-utilization function $H(x)$ and equals the sum of BIC and WIC [29]:

$$\text{TNW} = VAR[H(x)] = WIC + BIC \tag{4}$$

The metrics were originally developed for a single continuous variable x, released from any assumption of normality regarding the functional form of $p(y)$ and $\rho(x|y)$, and were subsequently implemented for the discrete case [52].

The ratio WIC/TNW was used to estimate the populations' degree of individual specialization [27,28]; a WIC/TNW close to one indicates a population of generalists, while a ratio close to zero indicates the dominance of specialist individuals.

Given that we had only one BIC, WIC, TNW and WIC/TNW value per combination population/season (see Results for further detail), we tested for differences in niche metrics by bootstrapping (see [53] for details). In brief, for each combination, we randomly sampled with replacement n individuals, where n corresponds to the sample size of the group. The procedure was repeated 999 times; the bootstrapped values had their distributions centred on the original values of the metrics, and differences were tested using a two-way ANOVA with season and population as orthogonal fixed factors. Similarly, Monte Carlo resampling simulations (9999 permutations) were used to evaluate if the degree of individual specialization WIC/TNW determined for each combination population/season was significantly different from that expected by a null scenario hypothesizing that all individuals sample equally from the population diet distribution. To test for correlations between pairwise size and diet similarity indices, a Mantel test with 9999 permutations was run.

A second index of individual diet specialisation was also considered; the individual specialisation (IS). This index was calculated as:

$$IS = \frac{\sum_i PS_i}{N} \tag{5}$$

where N is the number of i individuals in a population, while the proportional similarity index (PS_i) was estimated using the methodology proposed by Bolnick et al. [27]:

$$PS_i = 1 - 0.5 \sum_j \left| p_{ij} - q_j \right| \tag{6}$$

where p_{ij} is the frequency of a prey category j in the individual i's diet, and q_j is the frequency of prey category j in the entire population. We express values in the text as averages ± SE unless otherwise specified; for parametric statistical analysis, we tested data for conformity to the assumptions of variance homogeneity (Cochran's C test) and normality (Shapiro–Wilk test), and we log-transformed variables when required. All statistical analyses were performed in the R environment [54]. Specifically, the package *RInSp* (v. 1.2.3) was used for niche metrics analysis and related resampling procedures [53], *vegan* (v. 2.5-6; [55]) for nMDS and PERMANOVA analyses, and *ade4* (v. 1.7-13; [56]) for Mantel tests.

3. Results

At two sites (Populations 1 and 2) the proportion of arboreal vegetation cover was higher than the non-arboreal one, while in the third site (Population 3), the opposite occurred (Table 1). No significant difference in the arboreal cover was observed among the three sites; however, when considering the non-arboreal cover, Population 3 had a significantly higher proportion compared to the other two (Table 1).

Table 1. Location of the three *Hydromantes ambrosii* populations analysed in the study. Latitudes and longitudes are reported with a downgraded precision to increase species protection [46]. Information on elevation (m a.s.l.) are included, together with estimations of arboreal and non-arboreal vegetation cover (expressed in %) of the cave's surrounding areas. Vegetation variables were extracted from Terra MODIS VCF tiles (see text for further details) and, in the table, are averaged over the years 2016 and 2017 (SE in brackets). F values refer to the results of one-way ANOVAs testing for differences among the three locations, considering the years 2016 and 2017 as replicates. * = $p < 0.05$. The results of post-hoc Tukey HSD tests are also reported.

	Population 1	Population 2	Population 3	$F_{2,3}$	HSD
Longitude	9.77	9.72	9.72		
Latitude	44.12	44.18	44.17		
Elevation	331	206	260		
Arboreal vegetation	55.3 (3.6)	53.2 (3.5)	45.75 (1.85)	2.63	1 = 2 = 3
Non-arboreal vegetation	38 (0.4)	34.1 (2.6)	47.1 (1.4)	14.58 *	1 = 2 < 3

The nMDS plot of salamanders' stomach content data highlighted a considerable similarity in the dietary habits among the three populations of *Hydromantes ambrosii* (Figure 2); in contrast, independently from the population, salamanders showed a remarkably different diet when spring and fall were compared (Figure 2). PERMANOVA analyses confirmed that "season" significantly influenced salamanders' dietary habits, while negligible differences were found between populations, between ontogenetic stages or sexes, or between individuals of different size (Table 2).

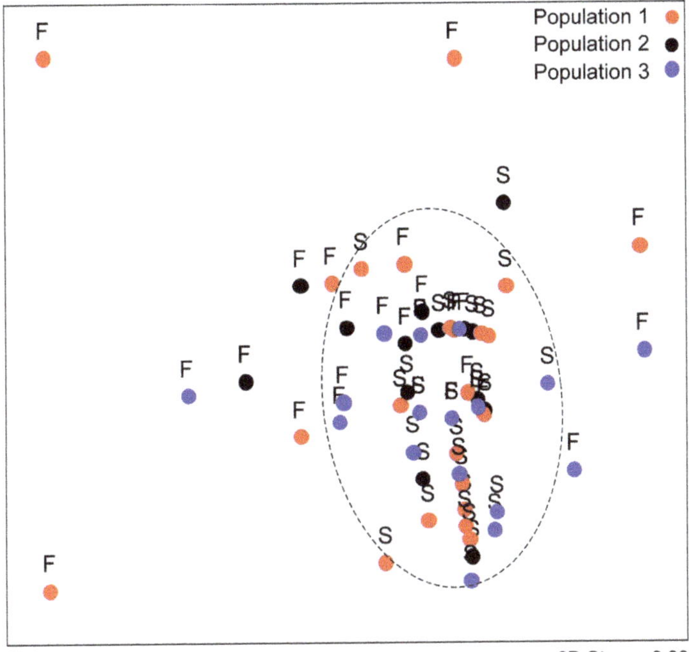

Figure 2. Non-metric multidimensional scaling plots (nMDS) of stomach content data of the three *H. ambrosii* populations. Letters indicate seasons (F = fall, S = spring). A 95% confidence ellipse is plotted around the spring data.

Table 2. Seasonal, age/sex and body size-related effects on stomach content data values across the three analysed *Hydromantes ambrosii* populations: results of PERMANOVA run on Euclidean distances calculated on log-transformed data and performed under a reduced model using 9999 permutations. p(MC) = p-value obtained by the Monte Carlo permutation test. The table reports the significance of factors, also testing pairwise interactions. On the left side are shown results considering all salamanders, and on the right side, those considering adults only; factor three represents age for all salamanders and sex for adults only. nd = factor or interaction not tested due to unbalanced design; SVL = snout-vent length.

Factor	All Salamanders		Adult Only	
	Pseudo-F	p(MC)	Pseudo-F	p(MC)
(1) SVL	0.91	0.454	0.54	0.781
(2) Population	1.23	0.246	0.99	0.444
(3) Age/Sex	1.5	0.175	1.5	0.164
(4) Season	16.03	0.001	14.07	0.001
(1) × (2)	1.14	0.328	1.16	0.28
(1) × (3)	1.64	0.124	0.29	0.976
(1) × (4)	0.77	0.570	1.14	0.308
(2) × (3)	0.44	0.981	0.70	0.755
(2) × (3)	0.67	0.788	0.75	0.702
(3) × (4)	1.25	0.257	1.92	0.077
(1) × (2) × (3)	1.25	0.258	0.95	0.469
(1) × (2) × (4)	1.65	0.067	1.45	0.142
(1) × (3) × (4)	0.48	0.846	2.02	0.045
(2) × (3) × (4)	nd	-	1.82	0.19
(1) × (2) × (3) × (4)	nd	-	nd	-

Niche metric analysis provided a more advanced resolution of the trophic characteristics of the three populations. In fall, BIC, WIC and TNW were approximately twice as large as in spring, independently of population (Table 3; two-way ANOVA, factor "season", $p < 0.05$ for both metrics), yet none showed seasonally consistent patterns of variation among populations (for all metrics, p always < 0.001 for the interaction factor "season $\times$ population"). In particular, in spring, Population 3 had the highest WIC followed by Populations 1 and 2, and BIC decreased from Population 1 to 3, while Population 1 showed the widest niche followed by Populations 3 and 2 (Table 3; post-hoc Tukey HSD tests, p always < 0.05). Conversely, in fall, WIC increased from Population 1 to 3, and Population 2 showed the highest BIC compared to the other two populations (HSD tests, p always < 0.05), while Population 1 was characterised by the narrowest niche compared to those of Populations 2 and 3 (HSD tests, $p < 0.05$), which show similar widths (HSD test, $p = 0.09$). The WIC/TNW ratio was characterised by a more consistent pattern among seasons (Figure 3). Independently from the season, salamanders from Populations 2 and 3 showed the highest and lowest degrees of individual specialization, respectively (HSD tests, p always < 0.01). Additionally, only Population 2 had, in both seasons, a WIC/TNW ratio significantly different from that expected by a null scenario, where all individuals sample equally from the pool of prey items of the population. Population 1 was characterised by an intermediate level of specialization, even though it was not significantly different from that of Population 2 in spring (HSD test, $p = 0.11$). The individual specialisation index (IS) was similar among the three populations and indicated a general higher proportion of specialised individuals in fall; however, the se indices were not statistically significant (Table 3).

The difference observed in the salamanders' condition index (BCI) among the studied populations was quite evident (Figure 4). In spring, individuals from Population 3 showed a low dietary specialization and were characterised by the highest BCI; contrarily, although Populations 1 and 2 had similar higher proportions of specialised individuals, the ir BCI dramatically differed (Figure 4). Interestingly, the patterns observed for the BCI values of the three populations partially replicated those of the vegetation characteristics of the cave surroundings: thus, Population 2 was associated with conditions of minimum non-arboreal and maximum arboreal cover, while the vegetation characteristics of the Population 3 cave surroundings showed the opposite (Table 1).

Table 3. Summary of trophic niche metrics and number of analysed individuals per season (in square brackets) of the three *Hydromantes ambrosii* populations. WIC = within-individual component; BIC = between-individual component; TNW = total niche width; IS = individual specialisation. 95% bootstrapped confidence intervals are in round brackets. The estimated density (N individuals/m^2) and, within parentheses, the total population size estimated by Ficetola et al. [48] of each population are included. [†] $p > 0.20$.

Season		Population 1	Population 2	Population 3
Fall		[3 F, 6 J]	[12 F, 4 M]	[4 F, 4 M, 2 J]
	WIC	1.394 (1.38–1.408)	1.482 (1.471–1.493)	1.729 (1.718–1.741)
	BIC	0.859 (0.853–0.866)	1.126 (1.118–1.135)	0.864 (0.858–0.87)
	TNW	2.253 (2.24–2.267)	2.608 (2.598–2.618)	2.594 (2.581–2.606)
	IS	0.442 [†]	0.340 [†]	0.427 [†]
Spring		[22 F, 14 M, 6 J]	[17 M, 15 M, 1 J]	[9 F, 5 M]
	WIC	0.684 (0.681–0.686)	0.603 (0.6–0.606)	0.799 (0.795–0.802)
	BIC	0.517 (0.509–0.524)	0.451 (0.444–0.458)	0.338 (0.334–0.343)
	TNW	1.201 (1.192–1.209)	1.054 (1.046–1.062)	1.137 (1.129–1.145)
	IS	0.728 [†]	0.699 [†]	0.738 [†]
Density		0.422 (144.5)	0.144 (137.2)	0.106 (67.1)

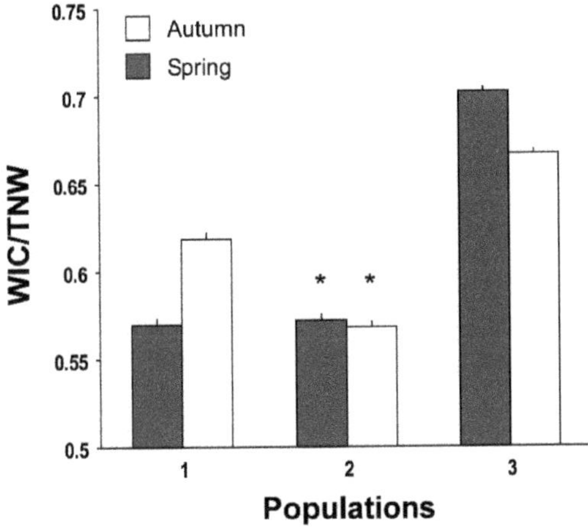

Figure 3. The ratio Within-Individual Component/Total Niche Width. Asterisks indicate that the ratios are significantly different from those expected by a null scenario hypothesising that all individuals sample equally from the population pool of prey items.

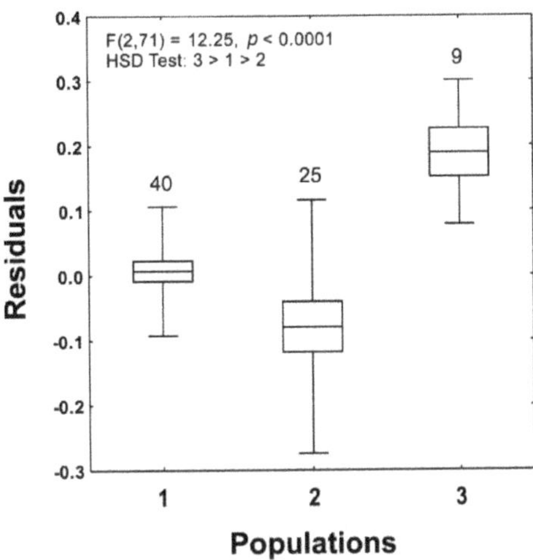

Figure 4. Variation in body condition across the three *H. ambrosii* populations. Boxes represent the ranges between first and third quartiles, horizontal bars inside the boxes represents the means, and whiskers embrace the full spectrum of data. Numbers on the tops of the whiskers represent the number of analysed salamanders. Top left: the results of one-way ANOVA and Tukey HSD tests assessing the significant difference among populations.

4. Discussion

Overall, in the present study, conventional multivariate statistical analyses indicated that the prey items consumed by the three *Hydromantes ambrosii* populations showed significant seasonal variations; however, negligible among-population effects were detected, suggesting that *H. ambrosii* salamanders may be characterised by dietary habits that are the same across the different localities. However, the analysis of the trophic niche metrics confirmed the importance of seasonality in the dietary habits of these salamanders, providing a far more advanced resolution of the actual among-population differences.

The total niche width (TNW) was about twice as wide in fall than in spring for all the three populations (Figure 2), confirming observations on other *Hydromantes* species [25,35]. The increase in the populations' trophic niche likely reflects a parallel increase in the diversity and availability of prey; indeed, the majority of the invertebrate taxa consumed by *Hydromantes* hibernate during winter and reproduce during spring/summer [57]. Thus, in fall, the re is probably a higher invertebrate diversity with the co-occurrence of different life stages, a condition increasing prey availability for *Hydromantes* salamanders [15,38]. The effect of seasonality on TNW was highly population-specific, with Population 2 showing the strongest variation, followed by Populations 3 and 1; similarly, both the intra- and inter-individual variation in the consumed prey items (WIC and BIC) varied significantly and unpredictably among populations and between seasons (Table 3). However, WIC was always higher than BIC (Table 2), indicating the prevalence of generalist individuals independently from the population. This was confirmed by the generally high values of the individual specialization index (WIC/TNW); however, only for Population 2 the comparison of the index against a null expectation was statistically significant in both spring and fall (Figure 3), indicating that this population may include "true specialist" individuals that do not randomly prey upon the available prey. A general lack of "true specialists" was also confirmed by the individual specialisation index (IS), which was never significantly different from the null hypothesis (Table 3). Only salamanders from Population 2 maintain a well-defined trophic strategy independently from seasonal variations in biotic constraints determined by, for example, density- or resource-dependent factors [26,38]. The se results are quite interesting as in the other six *Hydromantes* species for which dietary specialization was assessed [35–37]; all populations showed a significant large proportion of "true specialists". Conversely, in *H. ambrosii*, all populations were mostly composed of generalist individuals, and only one had a significant proportion of specialists (Figure 3). It is not clear if such a dominant generalism only characterises *H. ambrosii*, or if it is the result of particular environmental conditions acting locally. Future studies are needed to shed light on this particular topic.

The pattern of variation observed for the WIC/TNW ratio in spring (Figure 3) needs to be discussed considering the densities and body condition indices of the three populations, as well as differences in environmental conditions between the sites. Individuals from Population 2 showed a low body condition index and a relatively high degree of specialization (for both WIC/TNW and IS), while Population 3 showed a completely opposite pattern. Generally, when the density of a population is high, limited resources increase negative intraspecific interactions, and the energy consumption induced by competition leads to a general reduction in the body conditions of individuals [58,59]. In our case, Populations 2 and 3 showed the lowest and highest body conditions, respectively (Figure 4), although had similar estimated densities (Table 3; 0.14 vs. 0.11 individuals/m^2). One possible explanation for this inconsistency may rely be a difference in ecological opportunity characterising the outdoor surroundings of the caves. *Hydromantes* are not strictly cave species and forage close to the cave's entrance or in the surroundings, where prey are more abundant [8,21]. There, ecological opportunity (i.e., local prey abundance and diversity; [38]) is likely to be correlated with vegetation cover, the latter providing shelter, food resources and microclimatic conditions suitable for a multitude of invertebrate species [60,61] as well as salamanders. The arboreal vegetation showed limited variation across the three study sites; nevertheless, the outdoor surroundings of Population 2 were characterised by non-arboreal vegetation cover significantly lower than that of Population 3 (Table 1). This may also reflect a lower ecological opportunity in terms of prey abundance and, in turn, salamanders' body condition indices. Alternatively, the slightly lower arboreal vegetation cover characterising Population 3 might limit the microclimate suitable for invertebrates during dry periods, and these species likely seek shelter in the first few meters inside caves [8]. The refore, *Hydromantes* do not have to exit their refuge to prey, thus saving energy and reducing the stress due to less suitable environmental conditions and higher predation risks [10,22]. The se interesting hypotheses need to be tested as no data on prey abundance and diversity were available.

Negative biotic interactions (interspecific or intraspecific) are further processes potentially increasing the number of specialized individuals within a population, as those characterised by the weakest competitive abilities may reduce the width of their trophic niches as they only focus on sub-optimal food resources to avoid competition [32,62,63]. Interspecific competition is probably not playing an important role, since all *Hydromantes* species show allopatric distributions, and thus no more than one species occurs in the same area [2]. The only exception is the narrow contact zone between *H. ambrosii* and *H. italicus*, where hybrid populations occur [4]; however, the studied populations are located far from this hybrid zone. Moreover, no other Caudata are known to regularly exploit subaerial cave habitats, and thus competition with equivalent species can be ruled out. Competition might occur with some gecko species, as these reptiles can inhabit the areas close to the cave entrance [64]; still, geckos generally select dry microhabitats while *Hydromantes* select humid microhabitats [5], thus these species can hardly compete. Furthermore, interspecific competition with some cave invertebrates, although negligible and mostly affecting juvenile salamanders, cannot be excluded [11]. On the other hand, the occurrence of intraspecific competition in *Hydromantes* has not been yet assessed and the few available pieces of information are contrasting [65,66]. *Hydromantes* salamanders are facultative cave species seeking subterranean environments as a suitable place to escape the adverse climatic conditions occurring outside of the cave (too hot or dry) [2]. These species show a narrow microclimatic niche and select only cave areas falling within their tolerance range [5,9]. Thus, only specific areas of the caves can be occupied, a condition potentially leading to high densities [2,21]. However, we did not find signs of negative intraspecific competition. Indeed, during spring season, Population 1 had a much better BCI than Population 2 (Figure 4), although the former has a density of salamanders around three times higher, and the two have basically the same WIC/TNW ratio (Table 3). On the other hand, Population 3, although having a density comparable to that of Population 2 (Table 3), had higher a proportion of generalist individuals and BCI. The picture described here is drawn with information related to spring only, a season in which *Hydromantes* are extremely active [9,21], but we do not know if this pattern remains the same in fall. Therefore, further studies involving different species and seasons will help in confirming the lack of intraspecific competition observed here within the whole genus.

Noticeably, Population 2 showed the lowest mean BCI with the highest variance, suggesting that further factors may affect salamanders' conditions. This population inhabits a touristic cave; thus, disturbance produced by humans may cause stress to salamanders, altering their activity, with negative cascade effects on individuals' body conditions [21,67,68]. Indeed, an increase in salamanders' activity was observed in *H. flavus* (Sardinia, Italy) after gate-limited touristic activities in a cave (Manuela Mulargia, personal communication). Although specific data are not yet available, future studies are needed to confirm this hypothesis.

Nevertheless, our study has some limitations. The limited sample size of the dataset did not allow the assessment of the potential variation in the seasonal populations' niche breath across different years. To keep the highest possible number of analysed individuals, we assumed that the seasonal populations' niche breath did not significantly change among different years. This assumption deserves further confirmation through the analysis of multiple individuals collected over several years. Finally, two surveyed sites were spatially close (<180 m), and vegetation cover was estimated using a radius of 125 m around each site; thus, buffers partially overlapped. Although being partially non-independent, the differences in vegetation between these sites remained appreciable.

In conclusion, the present study provides new information on the dietary habits of *H. ambrosii*. Contrarily to most of the previous studies, in the present one, we focused on single populations to assess potential divergences occurring among conspecific populations. Such a kind of fine scale analysis highlighted significant variation in the trophic niche and degree of individual specialization among conspecific populations living close to each other. This helps to identify the potential drivers of divergences among populations' dynamics, therefore providing information to guide the implementation of ecologically meaningful conservation plans. Furthermore, the preponderance of generalist individuals observed among the studied populations is clearly in contrast with what already observed for other congeneric species [35–37]; this scenario deserves to be further investigated. This will necessarily include, in addition to the classic analysis of the stomach contents, other complementary methodologies such as stable isotope analysis [69,70], which will provide more complete information on the trophic ecology of this taxon.

Author Contributions: E.L. and G.M. conceived the idea, analyzed the data, wrote the first draft of the manuscript and prepared tables and figures; All authors have read and agreed to the published version of the manuscript.

Funding: Grant (NSFC-31972868) from the National Natural Science Foundation of China to Yahui Zhao.

Acknowledgments: We thank Sebastiano Salvidio for the invitation to contribute to this special issue. Fieldwork was authorized by the Italian Ministry of Environment (Prot. 9384/PNM-12/05/2015 and Prot. N. 20624/PNM-30/09/2016). E.L. is supported by the Chinese Academy of Sciences President's International Fellowship Initiative for postdoctoral researchers (2019PB0143). The final version of this paper has been concluded during a sabbatical year of one the authors (G.M.) in the Department of Chemistry, Biology, and Biotechnology, University of Perugia (Italy), and is dedicated to Sofia Mancinelli, thy eternal summer shall not fade.

Conflicts of Interest: The authors declare no conflicts of interest. The funder had no role in the design of the study; in the collection, analyses, or interpretation of data; in the writing of the manuscript, or in the decision to publish the results.

References

1. Wake, D.B. The enigmatic history of the European, Asian and American plethodontid salamanders. *Amphib. Reptil.* **2013**, *34*, 323–336. [CrossRef]

2. Lanza, B.; Pastorelli, C.; Laghi, P.; Cimmaruta, R. A review of systematics, taxonomy, genetics, biogeography and natural history of the genus *Speleomantes* Dubois, 1984 (Amphibia Caudata Plethodontidae). *Atti. Mus. Civ. Stor. Nat. Trieste* **2006**, *52*, 5–135.

3. Chiari, Y.; van der Meijden, A.; Mucedda, M.; Lourenço, J.M.; Hochkirch, A.; Veith, M. Phylogeography of Sardinian cave salamanders (genus *Hydromantes*) is mainly determined by geomorphology. *PLoS ONE* **2012**, *7*, e32332. [CrossRef] [PubMed]

4. Ficetola, G.F.; Lunghi, E.; Cimmaruta, R.; Manenti, R. Transgressive niche across a salamander hybrid zone revealed by microhabitat analyses. *J. Biogeogr.* **2019**, *46*, 1342–1354. [CrossRef]

5. Ficetola, G.F.; Lunghi, E.; Canedoli, C.; Padoa-Schioppa, E.; Pennati, R.; Manenti, R. Differences between microhabitat and broad-scale patterns of niche evolution in terrestrial salamanders. *Sci. Rep.* **2018**, *8*, 10575. [CrossRef] [PubMed]

6. Bruce, R.C.; Jaeger, R.G.; Houck, L.D. (Eds.) *The Biology of Plethodontid Salamanders*; Springer Science+Business Media, LLC: New York, NY, USA, 2000. [CrossRef]

7. Spotila, J.R. Role of temperature and water in the ecology of lungless salamanders. *Ecol. Monogr.* **1972**, *42*, 95–125. [CrossRef]

8. Culver, D.C.; Pipan, T. (Eds.) *The Biology of Caves and Other Subterranean Habitats*, 2nd ed.; Oxford University Press: New York, NY, USA, 2019.

9. Lunghi, E.; Manenti, R.; Ficetola, G.F. Seasonal variation in microhabitat of salamanders: Environmental variation or shift of habitat selection? *PeerJ* **2015**, *3*, e1122. [CrossRef]

10. Salvidio, S.; Palumbi, G.; Romano, A.; Costa, A. Safe caves and dangerous forests? Predation risk may contribute to salamander colonization of subterranean habitats. *Sci. Nat.* **2017**, *104*, 20. [CrossRef]

11. Pastorelli, C.; Laghi, P. Predation of *Speleomantes italicus* (Amphibia: Caudata: Plethodontidae) by *Meta menardi* (Arachnida: Araneae: Metidae). In *Atti del 6° Congresso Nazionale della Societas Herpetologica Italica (Roma, 27.IX-1.X.2006)*; Museo Regionale di Scienze Naturali: Torino, Italy, 2006; pp. 45–48.

12. Manenti, R.; Lunghi, E.; Canedoli, C.; Bonaccorsi, M.; Ficetola, G.F. Parasitism of the leech, *Batracobdella algira* (Moquin-Tandon, 1846), on Sardinian cave salamanders (genus *Hydromantes*) (Caudata: Plethodontidae). *Herpetozoa* **2016**, *29*, 27–35.

13. Lunghi, E.; Corti, C.; Manenti, R.; Barzaghi, B.; Buschettu, S.; Canedoli, C.; Cogoni, R.; De Falco, G.; Fais, F.; Manca, A.; et al. Comparative reproductive biology of European cave salamanders (genus *Hydromantes*): Nesting selection and multiple annual breeding. *Salamandra* **2018**, *54*, 101–108.

14. Deban, S.M.; Dicke, U. Motor control of tongue movement during prey capture in Plethodontid salamanders. *J. Exp. Biol.* **1999**, *202*, 3699–3714. [PubMed]

15. Lunghi, E.; Cianferoni, F.; Ceccolini, F.; Mulargia, M.; Cogoni, R.; Barzaghi, B.; Cornago, L.; Avitabile, D.; Veith, M.; Manenti, R.; et al. Field-recorded data on the diet of six species of European *Hydromantes* cave salamanders. *Sci. Data* **2018**, *5*, 180083. [CrossRef] [PubMed]

16. Deban, S.M.; Dicke, U. Activation patterns of the tongue-projector muscle during feeding in the imperial cave salamander *Hydromantes imperialis*. *J. Exp. Biol.* **2004**, *207*, 2071–2081. [CrossRef] [PubMed]

17. Deban, S.M.; Richardson, J.C. Cold-Blooded snipers: Thermal independence of ballistic tongue projection in the salamander *Hydromantes platycephalus*. *J. Exp. Zool.* **2011**, *315*, 618–630. [CrossRef] [PubMed]

18. Vignoli, L.; Caldera, F.; Bologna, M.A. Trophic niche of cave populations of *Speleomantes italicus*. *J. Nat. Hist.* **2006**, *40*, 1841–1850. [CrossRef]

19. Salvidio, S.; Romano, A.; Oneto, F.; Ottonello, D.; Michelon, R. Different season, different strategies: Feeding ecology of two syntopic forest-dwelling salamanders. *Acta Oecol.* **2012**, *43*, 42–50.

20. Salvidio, S. Diet and food utilization in the European plethodontid *Speleomantes ambrosii*. *Vie et Milieu* **1992**, *42*, 35–39.

21. Lunghi, E.; Manenti, R.; Mulargia, M.; Veith, M.; Corti, C.; Ficetola, G.F. Environmental suitability models predict population density, performance and body condition for microendemic salamanders. *Sci. Rep.* **2018**, *8*, 7527. [CrossRef]

22. Jiménez-Valverde, A.; Sendra, A.; Garay, P.; Reboleira, A.S.P.S. Energy and speleogenesis: Key determinants of terrestrial species richness in caves. *Ecol. Evol.* **2017**, *7*, 10207–10215. [CrossRef]

23. Crump, M.L. Intra-population variability in energy parameters of the salamander *Plethodon cinereus*. *Oecologia* **1979**, *38*, 235–247. [CrossRef]

24. Marvin, G.A.; Bryan, R.; Hardwick, J. Effect of chronic low body temperature on feeding and gut passage in a plethodontid salamander. *J. Therm. Biol.* **2017**, *69*, 319–324. [CrossRef] [PubMed]

25. Lunghi, E.; Cianferoni, F.; Ceccolini, F.; Veith, M.; Manenti, R.; Mancinelli, G.; Corti, C.; Ficetola, G.F. What shapes the trophic niche of European plethodontid salamanders? *PLoS ONE* **2018**, *13*, e0205672. [CrossRef] [PubMed]

26. Bolnick, D.I.; Svanbäck, R.; Fordyce, J.A.; Yang, L.H.; Davis, J.M.; Hulsey, C.D.; Forister, M.L. The ecology of individuals: Incidence and implications of individual specialization. *Am. Nat.* **2003**, *161*, 1–28. [CrossRef] [PubMed]

27. Bolnick, D.I.; Yang, L.H.; Fordyce, J.A.; Davis, J.M.; Svanbäck, R. Measuring individual-level resource specialization. *Ecology* **2002**, *83*, 2936–2941. [CrossRef]

28. Roughgarden, J. Evolution of niche width. *Am. Nat.* **1972**, *106*, 683–718. [CrossRef]

29. Roughgarden, J. Niche width: Biogeographic patterns among *Anolis* lizard populations. *Am. Nat.* **1974**, *108*, 429–442. [CrossRef]

30. Bolnick, D.I.; Amarasekare, P.; Araújo, C.S.; Bürger, R.; Levine, J.M.; Novak, M.; Rudolf, V.H.W.; Schreiber, S.J.; Urban, M.C.; Vasseur, D.A. Why intraspecific trait variation matters in community ecology. *Trends Ecol. Evol.* **2011**, *26*, 183–192. [CrossRef]

31. Layman, C.A.; Newsome, S.D.; Crawford, T.G. Individual-level niche specialization within populations: Emerging areas of study. *Oecologia* **2015**, *178*, 1–4. [CrossRef]

32. Svanbäck, R.; Bolnick, D.I. Intraspecific competition drives increased resource use diversity within a natural population. *Proc. R. Soc. B* **2007**, *274*, 839–844. [CrossRef]

33. Araújo, M.S.; Bolnick, D.I.; Martinelli, L.A.; Giaretta, A.A.; dos Reis, S.F. Individual-level diet variation in four species of Brazilian frogs. *J. Anim. Ecol.* **2009**, *78*, 848–856. [CrossRef]

34. Terraube, J.; Guixé, D.; Arroyo, B. Diet composition and foraging success in generalist predators: Are specialist individuals better foragers? *Basic Appl. Ecol.* **2014**, *15*, 616–624. [CrossRef]

35. Salvidio, S.; Oneto, F.; Ottonello, D.; Costa, A.; Romano, A. Trophic specialization at the individual level in a terrestrial generalist salamander. *Can. J. Zool.* **2015**, *93*, 79–83. [CrossRef]

36. Salvidio, S.; Pasmans, F.; Bogaerts, S.; Martel, A.; van de Loo, M.; Romano, A. Consistency in trophic strategies between populations of the Sardinian endemic salamander *Speleomantes imperialis*. *Anim. Biol.* **2017**, *67*, 1–16. [CrossRef]

37. Lunghi, E.; Manenti, R.; Cianferoni, F.; Ceccolini, F.; Veith, M.; Corti, C.; Ficetola, G.F.; Mancinelli, G. Inter-specific and inter-population variation in individual diet specialization: Do environmental factors have a role? *Ecology* **2020**. [CrossRef]

38. Araújo, M.S.; Bolnick, D.L.; Layman, C.A. The ecological causes of individual specialisation. *Ecol. Lett.* **2011**, *14*, 948–958. [CrossRef]

39. Araujo, M.S.; dos Reis, S.F.; Giaretta, A.A.; Machado, G.; Bolnick, D.I. Intrapopulation diet variation in four frogs (*Leptodactylidae*) of the Brazilian savannah. *Copeia* **2007**, *2007*, 855–865. [CrossRef]

40. Pyke, G.H.; Pullian, H.R.; Charnov, E.L. Optimal foraging: A selective review of theory and tests. *Quart. Rev. Biol.* **1977**, *52*, 137–154. [CrossRef]

41. Quiroga, M.F.; Bonansea, M.I.; Vaira, M. Population diet variation and individual specialization in the poison toad, *Melanophryniscus rubriventris* (Vellard, 1947). *Amphib. Reptil.* **2011**, *32*, 261–265. [CrossRef]

42. Rosenblatt, A.E.; Nifong, J.C.; Heithaus, M.R.; Mazzotti, F.J.; Cherkiss, M.S.; Jeffery, B.M.; Elsey, R.M.; Decker, R.A.; Silliman, B.R.; Guillette, L.G.J.; et al. Factors affecting individual foraging specialization and temporal diet stability across the range of a large "generalist" apex predator. *Oecologia* **2015**, *178*, 5–16. [CrossRef]

43. Lunghi, E.; Bruni, G. Long-term reliability of Visual Implant Elastomers in the Italian cave salamander (*Hydromantes italicus*). *Salamandra* **2018**, *54*, 283–286.

44. Salvidio, S. Homing behaviour in *Speleomantes strinatii* (Amphibia Plethodontidae): A preliminary displacement experiment. *North West. J. Zool.* **2013**, *9*, 429–433.

45. Smith, M.A.; Green, D.M. Dispersal and the metapopulation paradigm in amphibian ecology and conservation: Are all amphibian populations metapopulations? *Ecography* **2005**, *28*, 110–128. [CrossRef]

46. Lunghi, E.; Corti, C.; Manenti, R.; Ficetola, G.F. Consider species specialism when publishing datasets. *Nat. Ecol. Evol.* **2019**, *3*, 319. [CrossRef] [PubMed]

47. Novak, M.; Tinker, M.T. Timescales alter the inferred strength and temporal consistency of intraspecific diet specialization. *Oecologia* **2015**, *178*, 61–74. [CrossRef] [PubMed]

48. Ficetola, G.F.; Barzaghi, B.; Melotto, A.; Muraro, M.; Lunghi, E.; Canedoli, C.; Lo Parrino, E.; Nanni, V.; Silva-Rocha, I.; Urso, A.; et al. N-mixture models reliably estimate the abundance of small vertebrates. *Sci. Rep.* **2018**, *8*, 10357. [CrossRef]

49. Băncilă, R.I.; Hartel, T.; Plăiaşu, R.; Smets, J. Cogălniceanu, D. Comparing three body condition indices in amphibians: A case study of yellow-bellied toad *Bombina variegata*. *Amphib. Reptil.* **2010**, *31*, 558–562. [CrossRef]

50. Labocha, M.K.; Schutz, H.; Hayes, J.P. Which body condition index is best? *Oikos* **2014**, *123*, 111–119. [CrossRef]

51. Scott, D.E.; Casey, E.D.; Donovan, M.F.; Lynch, T.K. Amphibian lipid levels at metamorphosis correlate to post-metamorphic terrestrial survival. *Oecologia* **2007**, *153*, 521–532. [CrossRef]

52. Roughgarden, J. *Theory of Population Genetics and Evolutionary Ecology*; Macmillan Publishing Company: New York, NY, USA, 1979; p. 612.

53. Zaccarelli, N.; Bolnick, D.I.; Mancinelli, G. RInSp: An R package for the analysis of individual specialization in resource use. *Methods Ecol. Evol.* **2013**, *4*, 1018–1023. [CrossRef]

54. *R: A Language and Environment for Statistical Computing*; R Foundation for Statistical Computing: Vienna, Austria. Available online: http://www.R-project.org/ (accessed on 18 September 2019).

55. Oksanen, J.; Blanchet, F.G.; Friendly, M.; Kindt, R.; Legendre, P.; McGlinn, D.; Minchin, P.R.; O'Hara, R.B.; Simpson, G.L.; Solymos, P.; et al. Vegan: Community Ecology Package. R Package Version 2.5-6. Available online: http://cran.r-project.org/web/packages/vegan (accessed on 28 January 2020).

56. Dray, S.; Dufour, A.B. ade4: Analysis of Ecological Data - Exploratory and Euclidean Methods in Environmental Sciences. Available online: https://CRAN.R-project.org/package=ade4 (accessed on 28 January 2020).

57. Bale, J.S.; Hayward, S.A.L. Insect overwintering in a changing climate. *J. Exp. Biol.* **2010**, *213*, 980–994. [CrossRef]

58. Kirk, D.A.; Gosler, A.G. Body condition varies with migration and competition in migrant and resident south American vultures. *Auk* **1994**, *111*, 933–944. [CrossRef]

59. Petren, K.; Case, T.J. Habitat structure determines competition intensity and invasion success in gecko lizards. *Proc. Natl. Acad. Sci. USA* **1998**, *95*, 11739–11744. [CrossRef] [PubMed]

60. Baur, B.; Cremene, C.; Groza, G.; Schileyko, A.A.; Baur, A.; Erhardt, A. Intensified grazing affects endemic plant and gastropod diversity in alpine grasslands of the Southern Carpathian mountains (Romania). *Biologia* **2007**, *62*, 438–445. [CrossRef]

61. Norbury, G.; Heyward, R.; Parkes, J. Skink and invertebrate abundance in relation to vegetation, rabbits and predators in a New Zealand dryland ecosystem. *N. Z. J. Ecol.* **2009**, *33*, 24–31.

62. Tinker, M.T.; Bentall, G.; Estes, J.A. Food limitation leads to behavioral diversification and dietary specialization in sea otters. *Proc. Natl. Acad. Sci. USA* **2008**, *105*, 560–565. [CrossRef] [PubMed]

63. Araújo, M.S.; Costa-Pereira, R. Latitudinal gradients in intraspecific ecological diversity. *Biol. Lett.* **2013**, *9*, 20130778. [CrossRef] [PubMed]

64. Nguyen, H.N.; Lu, C.-W.; Chu, J.-H.; Grismer, L.L.; Hung, C.-M.; Lin, S.-M. Historical demography of four gecko species specializing in boulder cave habitat – its implications in the evolutionary dead end hypothesis and conservation. *Mol. Ecol.* **2019**, *28*, 772–784. [CrossRef] [PubMed]

65. Salvidio, S.; Pastorino, M.V. Spatial segregation in the European plethodontid *Speleomantes strinatii* in relation to age and sex. *Amphib. Reptil.* **2002**, *23*, 505–510.

66. Ficetola, G.F.; Pennati, R.; Manenti, R. Spatial segregation among age classes in cave salamanders: Habitat selection or social interactions? *Pop. Ecol.* **2013**, *55*, 217–226. [CrossRef]

67. Cejuela Tanalgo, K.; Tabora, J.A.G.; Hughes, A.C. Bat cave vulnerability index (BCVI): A holistic rapid assessment tool to identify priorities for effective cave conservation in the tropics. *Ecol. Indic.* **2018**, *89*, 852–860. [CrossRef]

68. Whitten, T. Applying ecology for cave management in China and neighbouring countries. *J. Appl. Ecol.* **2009**, *46*, 520–523. [CrossRef]

69. Cloyed, C.S.; Eason, P.K. Niche partitioning and the role of intraspecific niche variation in structuring a guild of generalist anurans. *R. Soc. Open Sci.* **2017**, *4*, 170060. [CrossRef] [PubMed]

70. Costa-Pereira, R.; Rudolf, V.H.W.; Souza, F.L.; Araújo, M.S. Drivers of individual niche variation in coexisting species. *J. Anim. Ecol.* **2018**, *87*, 1452–1464. [CrossRef] [PubMed]

Article

Do Salamanders Limit the Abundance of Groundwater Invertebrates in Subterranean Habitats?

Raoul Manenti [1,2,*], Enrico Lunghi [3,4], Benedetta Barzaghi [1], Andrea Melotto [5], Mattia Falaschi [1] and Gentile Francesco Ficetola [1,6]

[1] Department of Environmental Science and Policy, Università degli Studi di Milano, via Celoria, 26, 20133 Milano, Italy; benedetta.barzaghi@unimi.it (B.B.); mattia.falaschi@unimi.it (M.F.); francesco.ficetola@unimi.it (G.F.F.)
[2] Laboratorio di Biologia Sotterranea "Enrico Pezzoli", Parco Regionale del Monte Barro, 23851 Galbiate, Italy
[3] Key Laboratory of the Zoological Systematics and Evolution, Institute of Zoology, Chinese Academy of Sciences, Beichen West Road 1, Beijing 100101, China; enrico.arti@gmail.com
[4] Museo di Storia Naturale dell'Università degli Studi di Firenze, sede "La Specola", Via Romana 17, 50125 Firenze, Italy
[5] Centre for Invasion Biology, Stellenbosch University, Matieland 7602, South Africa; mel8@hotmail.it
[6] LECA, Laboratoire d'Ecologie Alpine, CNRS, The Grenoble Alpes University and The Savoie Mont Blanc University, F-38000 Grenoble, France
* Correspondence: raoulmanenti@gmail.com

Received: 19 March 2020; Accepted: 15 April 2020; Published: 20 April 2020

Abstract: Several species of surface salamanders exploit underground environments; in Europe, one of the most common is the fire salamander (*Salamandra salamandra*). In this study, we investigated if fire salamander larvae occurring in groundwater habitats can affect the abundance of some cave-adapted species. We analyzed the data of abundance of three target taxa (genera *Niphargus* (Amphipoda; Niphargidae), *Monolistra* (Isopoda; Sphaeromatidae) and *Dendrocoelum* (Tricladida; Dedrocoelidae)) collected in 386 surveys performed on 117 sites (pools and distinct subterranean stream sectors), within 17 natural and 24 artificial subterranean habitats, between 2012 and 2019. Generalized linear mixed models were used to assess the relationship between target taxa abundance, fire salamander larvae occurrence, and environmental features. The presence of salamander larvae negatively affected the abundance of all the target taxa. *Monolistra* abundance was positively related with the distance from the cave entrance of the sites and by their surface. Our study revealed that surface salamanders may have a negative effect on the abundance of cave-adapted animals, and highlited the importance of further investigations on the diet and on the top-down effects of salamanders on the subterranean communities.

Keywords: cave biology; prey; hypogean; underground; stygofauna; *Monolistra*; Sphaeromatidae; *Niphargus*; flatworm; aqueduct; seepage

---

1. Introduction

Salamanders represent an important fraction of aquatic and terrestrial biomass in several environments. Salamanders typically display a life cycle involving aquatic larvae and terrestrial adults. However, several peculiar adaptations to a total terrestrial or to a complete aquatic life evolved separately in different salamanders' lineages allowing the exploitation of a large variety of environments [1]. In both cases, salamanders often retain the role of keystone predators, affecting the structure of the communities in different aquatic and terrestrial habitats [2]. In temporary ponds, salamanders are known to regulate the nutrient flows within aquatic food webs by affecting the abundance of zooplankton and tadpoles [3]. In forests, salamanders are often abundant mesopredators that can strongly affect the abundance and composition of invertebrate communities, sometimes even mediating the rates of leaf litter decomposition [4]. Salamanders' predatory activity can also determine trophic cascades (such as changes in the trophic web across two or more links) especially in communities based on detritus [5]. For example, the red-backed salamander (*Plethodon cinereus*) is an abundant predator on springtails, mites, and other small prey invertebrates [6], which in turn feed upon a large fungal biomass [7]. Thus, red-backed salamanders play top-down effects on fungal communities of forest floors [8].

Among the environments with detritivore-based trophic webs, groundwaters provide a promising research field that deserves to be implemented for different reasons. First of all, groundwater represents the major source of potable water supply for humans and, globally, is the largest source of available freshwater [9,10]. Second, underground freshwater environments such as aquifers, hyporheic zones, and cave rivers can be of particular interest to understand processes shaping global biodiversity. In these environments, ecological variation is weak compared to surface habitats, and this facilitates studies assessing mechanisms that allow colonization by animals and the differentiation of colonizers that often follows [11,12]. Although the majority of studies on groundwater fauna report findings and descriptions of new species, there is an increasing interest in understanding the evolutionary processes involved in cave colonization and the distribution of cave adapted animals [13,14]. Most studies on animals inhabiting groundwater refer to "stygobionts", i.e., those animals that evolved specific adaptations to underground freshwater habitats, in which they spend their entire life-cycle [15]. Among them there are at least 13 species or subspecies of salamanders that are considered obligate cave-dwellers and display typical morphological adaptations (e.g., eyeless and depigmentation) to the subterranean environment [16]. These salamanders are often fully aquatic and occupy the top predator role in groundwaters. However, non-obligate cave-dwelling salamander species can also play a fundamental role in shaping cave food webs. Several species of surface salamanders are known to exploit underground environments where they can feed on invertebrates, guano, or other urodeles [17–20]. Moreover, some of them often breed in subterranean rivers and streams where their larvae are able to complete the entire lifecycle [21]. An increasing number of studies are showing their ecological and evolutionary importance. As an example, they can help us understand the dynamics of novel habitat colonizations and provide useful insights to understand the relative role of phenotypic plasticity and local adaptations [22]. Moreover, as they can reach high abundances and show well defined patterns (i.e., seasonal, ecological) of cave exploitations, they can exert important roles on the community inhabiting the surrounding of the cave entrance and the twilight zone [23,24].

In the last years, a growing number of studies has investigated the ecology of fire salamanders (*Salamandra salamandra*) breeding in underground environments. This species is an ovoviviparous widespread amphibian in Europe that shows high ecological plasticity in the choice of breeding sites [25–27]. This salamander can breed in numerous subterranean environments; larvae can be found in natural caves streams and pools, artificial hypogean springs and flooded mines where they may reach high densities [28]. Generally, in groundwaters, fire salamander larvae are found within the first 5–30 m from the cave entrance; however, records of larvae in deeper areas (>100 m) are also reported [21,29]. In most of the groundwater sites where fire salamander breeds, larvae occupy the top predator position [30,31], however, prey is often rare and food scarcity poses major constraints to their development [32].

An aspect that is still not well understood is the role played by fire salamander larvae on stygobiont fauna. The exploitation of groundwater by animals normally occurring at the surface can determine changes in the composition of stygobiont communities [33]. For example, a recent study considering natural and artificial spring habitats revealed that the occurrence of fire salamander larvae limits the occurrence of the stygobiont *Niphargus thuringius* at the interface between groundwater and surface streams [34].

With this study we aim to assess if the occurrence of fire salamander larvae in subterranean habitats affects the abundance of cave adapted animals. We predict that, irrespective to the distance from the entrance, groundwater sites with fire salamander larvae show a lower abundance of stygobiont species than groundwater sites without larvae.

2. Materials and Methods

2.1. Sampling Design

Between 2012 and 2019, we performed repeated sampling of subterranean aquatic fauna across multiple caves and artificial subterranean habitats with streams or pools. Each considered site was visited at least twice during the same season of the same year; we only considered in analyses sites where the occupancy of fire salamander larvae did not change between surveys (i.e., salamander larvae were always present or absent during the successive surveys). The subterranean sites (Figure 1) are located between the districts of Como, Lecco, Bergamo, and Monza and Brianza of Lombardy and of La Spezia in Liguria (NW-Italy).

To obtain preliminary information on caves (i.e., location, development) we used the data from the cave cadasters of Lombardy and Liguria. The artificial subterranean sites considered here were artificial subterranean springs, draining galleries of catchment (the so called 'bottini'), and artificial mines. To localize the artificial subterranean sites, we used information available in studies on subterranean fauna [35] and local information on mine activity.

Visual encounter surveys were performed to assess salamander larvae occurrence and stygobiont abundance. Water depth and distance from the entrance were also measured. In all sites we sampled the largest pool or waterbody that we found and the other pools or streams that occurred. In streams, we randomly choose one or more sections of the watercourse from the cave entrance to the deepest part that we reached. Overall, we performed 386 visual samplings, surveying 117 sites (pools and distinct subterranean stream sectors) within 17 natural caves and 24 artificial cavities.

We searched both stygobiont fauna and fire salamander larvae by employing standardized visual encounter surveys, during which each pool or stream's sector was actively investigated with a constant effort of 3 min/m^2 [36]. The detection probability of fire salamander larvae is generally high, especially during nights in surface environments and in groundwaters and visual observations that allow to detect the species occurrence with confidence >0.95 [37], thus false absences in our analyses are unlikely.

We assessed the potential role of fire salamander larvae on the abundance of three taxa of stygofauna: amphipods of the genus *Niphargus*, isopods of the genus *Monolistra*, and planarians of the genus *Dendrocoelum* (Figure 2). All these animals show features typical of cave-adapted species, such as eyeless and depigmentation. *Monolistra* isopods generally feed on detritus and biofilms occurring on the substrate and composed of fungi and bacteria [38,39]. *Niphargus* crustaceans show a generalist diet comprising both plant debris and other arthropods, and display both a detritivore and a predatory/cannibalistic behavior [40,41]. Planarians are predators and can hold the highest position of the food web in small interstitial groundwater habitats or where salamanders do not occur. In this study, we focused at the genus level and, for each genus, we included multiple species in the analyses. Multiple *Niphargus* species are present in all the groundwaters of the study area but high confusion regarding their taxonomy exists [39]. In the study area, *Monolistra* crustaceans include different species that occupy different distinct karst areas; we investigated caves in the range of *M. pavani*, *M. bergomas*, and *M. julia*. Only a limited number of planarians of the genus *Dendrocoelum* are currently described for Italian caves [42], and during our investigations we have recorded a higher number of localities at which these flatworms occur.

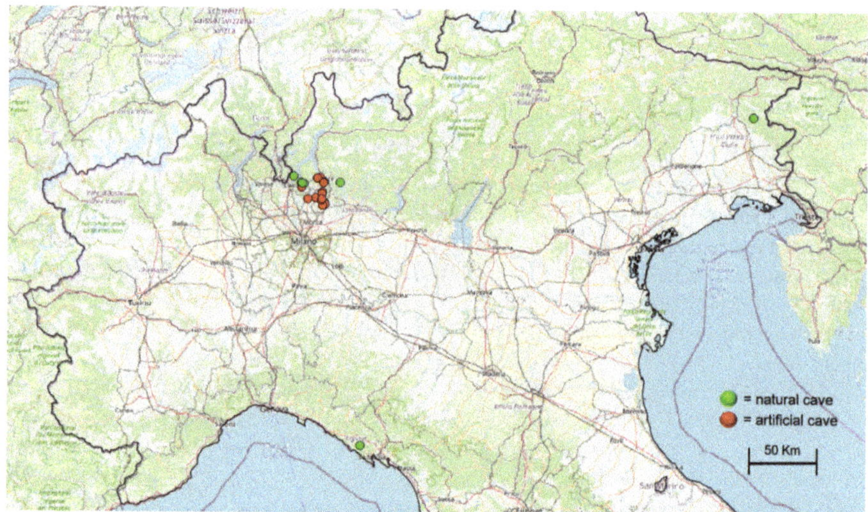

Figure 1. Sampling caves considered in this study. Caves are divided into natural (green dots) and artificial (red dots). Due to geographic proximity most of the sites are superimposed.

2.2. Statistical Analyses

We used generalized linear mixed models (GLMMs) to assess the relationships between the abundance of the target taxa, salamander occurrence, and habitat features. Generalized mixed models yield reliable estimates of the relationships between the relative abundance of animals and environmental conditions [43]. Before performing GLMMs we checked correlations between all the variables. We performed three distinct GLMMs, one for each stygobiont taxon. Sites outside the range of the *Monolistra* species were excluded from the analysis focusing on this genus (Supplementary Table S1), to avoid bias related to biogeographical patterns. As dependent variables, we considered the number of active individuals of the target taxa observed for each site at each sampling occasion. As an independent variable we used the occurrence of fire salamander larvae, the distance from the cave entrance, and the maximum water depth; we included also the area of the sites as covariate. As random factors we considered the cavity in which we sampled the subterranean pools or the streams and the year of sampling. We built models using negative binomial distribution (type I). For each GLMM we tested all combinations of explanatory variables for multicollinearity using the variance inflation factor (VIF); all VIF were below 2. We assessed significance of variables in GLMMs using a likelihood ratio test.

GLMMs were run in R environment (R Development Core Team 2018) using the packages lmerTest [44], glmmTMB [45], and car [46].

3. Results

Fire salamander larvae occurred in 41 sites (23 caves). The most widespread cave-dwelling taxon was the genus *Niphargus* that was detected in 48 sites (23 caves). By contrast, the planarians of the genus *Dendrocoelum* were more localized, occurring in 28 sites (11 caves), while crustaceans of the genus *Monolistra* were recorded in 25 sites (four caves only) and 47 sampling occasions (Table S1). Although localized, *Monolistra* reached the maximum abundance recorded at a single site with 106 individuals. Instead, maximum abundance at a single site was 56 individuals for *Dendrocoelum* and 20 individuals for *Niphargus*. The proportion of surveyed microhabitats occupied per cave varied consistently; generally, in the caves where we detected *Monolistra* occurrence, these isopods were detected in 90% of the sites, while *Niphargus* and *Dendrocoelum* were detected in a substantially minor fraction of microhabitats.

We detected *Monolistra* only in natural caves and in natural microhabitats, while both *Niphargus* and *Dendrocoelum* occurred also in different artificial pools of draining galleries.

GLMMs revealed that the occurrence of fire salamander larvae played a significant effect on the abundance of all the target taxa (Table 1). All target taxa showed a reduced abundance in sites with salamander larvae (Figure 3). The abundance of *Niphargus* was positively related to the maximum depth of the sites with higher densities occurring in deeper sites (Table 1). For *Monolistra* the analysis also revealed a tendency to occupy habitats farther from the entrance and with larger area (Table 1).

Figure 2. Examples of the taxa considered in the study: (**a**) *Salamandra salamandra* larvae at different stages; (**b**) a *Dendrocoelum* flatworm from the Pignone cave (Liguria); (**c**) an isopod crustacean of the genus *Monolistra* (*Monolistra pavani*); (**d**) an amphipod crustacean of the genus *Niphargus* (*Niphargus thuringius*). Credits R. Manenti.

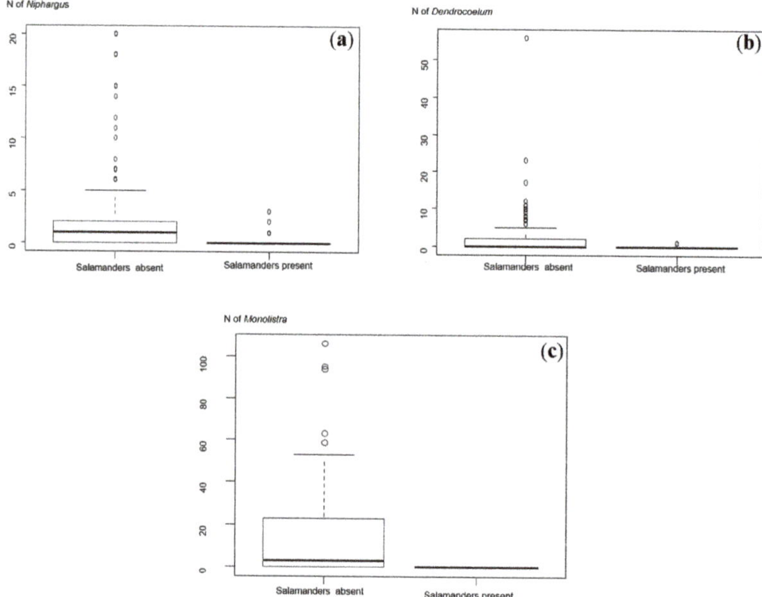

Figure 3. Boxplots of relationships between the occurrence of fire salamander larvae and abundance of the stygobiont target taxa: (**a**) amphipods of the genus *Niphargus*; (**b**) planarians of the genus *Dendrocoelum*; (**c**) isopods of the genus *Monolistra*.

Table 1. Results of the likelihood ratio test on generalized linear mixed models (GLMMs) assessing the relationship between the presence of fire salamander larvae and environmental variables with the abundance of the three target stygobiont taxa. Significant relationships are in bold.

	Variables	Estimate	SE	χ2	P
Niphargus					
	Fire salamander larvae	**−3.34**	**0.44**	**78.31**	**<0.001**
	Distance from surface	<0.01	<0.01	0.01	0.94
	Maximum water depth	**0.01**	**<0.01**	**4.29**	**0.03**
	Surveyed area	−0.05	0.05	1.043	0.30
Dendrocoelum					
	Fire salamander larvae	**−2.39**	**0.84**	**11.53**	**<0.01**
	Distance from surface	<0.01	<0.01	1.35	0.24
	Maximum water depth	<−0.01	0.01	0.57	0.44
	Surveyed area	0.06	0.07	0.74	0.38
Monolistra					
	Fire salamander larvae	**<−0.01**	**<0.01**	**6.24**	**0.01**
	Distance from surface	**<0.01**	**<0.01**	**4.71**	**0.02**
	Maximum water depth	<0.01	<0.01	0.67	0.41
	Surveyed area	**<0.01**	**<0.01**	**6.98**	**<0.01**

4. Discussion

This is the first study that investigated the relationship between facultative cave-breeding salamanders and the relative abundance of invertebrate fauna adapted to groundwaters. Our results indicate that the occurrence of fire salamander larvae in groundwaters may limit the density of different stygobiont animals such as crustaceans and planarians, showing that these animals can shape the diversity of fauna in groundwaters, at least nearby the surface. Previous studies have shown that salamander occurrence in caves is favored by some cave features, such as the stability of habitat (water permanence), the absence of predators, and the availability of resources [47,48]. Caves and other subterranean environments with groundwater may offer more stable breeding habitats, with a more regular hydroperiod, than surface streams and creeks, which especially in karst landscapes may be subjected to strong variation depending on the amount of rainfalls [26]. Moreover, cave pools are usually predator-deprived environments and can be considered as safe habitats for the fire salamander larvae [31]. However, these environments also harbor low densities of invertebrate prey, especially when compared to surface breeding sites, posing a constraint to larval development [49].

Fire salamander larvae are generalist predators that can prey upon a large range of invertebrates [50–52]. In groundwater habitats that are close to surface, animals from outside, like dipterans and crustaceans, may occur and become prey of fire salamander larvae [33,53]. However, stygobiont species can also constitute a useful resource and be opportunistically preyed. For these animals, the trophic perspective is reversed because, compared to deepest sectors, the underground habitats close to the surface can be richer in terms of available food [54–56]. Thus sectors close to the surface and surface habitats themselves can provide useful trophic resources for stygobionts which, in favorable seasons or with particular environmental conditions, can occupy springs or move closer to the cave entrance. At the same time, areas at the boundary between underground and surface environments can be more risky in terms of climatic variation (they are unstable compared to deep subterranean habitats) and predator occurrence [57,58]. Our results suggest that when a top-predator occurs in subterranean habitats, it may severely limit the abundance of stygobiont fauna, since all the three target taxa considered in this study showed a significant lower abundance in sites with fire salamander larvae. To assess the effect of direct predation by fire salamander larvae on stygofauna further investigations using stomach flushing or stable isotope analysis are necessary.

In particular, we observed a negative relationship between salamander larvae and the abundance of *Niphargus* crustaceans and *Dendrocoelum* planarians. With more than 430 described species at the global scale, *Niphargus* is the most diverse genus of freshwater amphipods [59–61]. It is widespread and primarily inhabits groundwaters, but several *Niphargus* species/populations live in subterranean habitats at the interface with the surface and can more or less occasionally exploit epigean environments like springs and streams [34,61,62]. Even if eyeless, *Niphargus* species retain the ability to detect light [63] suggesting that the connection with surface environments and exploitation of transitional habitats can be important for these crustaceans. Salamander larvae can exhert major predatory pressure on these animals; it is also important to consider that the occurrence of salamander larvae is seasonal and, even if their development can be quite long [64,65], periods in which larvae are absent are likely to occur. If we consider also that the biomass of laid larvae is generally higher than that of metamorphosing one [32], it is possible that the subsidization by fire salamander can also have effects on *Niphargus* and other organisms when the predation pressure is not present or present only in adjacent microhabitats. Further investigations could be performed by surveying the same subterranean habitats when there will be no fire salamander larvae inside. *Niphargus* abundance was also positively related to water depth. Other than hosting a higher water volume to be surveyed, deeper pools can provide more shelters during water flow and host more organic debris.

Dendrocoelum planarians are predators of annelid, crustaceans, and other invertebrates [66,67]; very few studies are available for subterranean species and there is lack of ecological information on factors favoring their abundance and distribution. Only a few cave species with very narrow ranges are currently known in Italy [42]. *Dendrocoelum* planarians, when fire salamander larvae occur, can be considered as mesopredator; our data suggests however that the effect of fire salamander larvae is similar in planarians, *Niphargus*, and detritivore *Monolistra* as well. Predation of salamanders on subterranean planarians has been observed in the case of the Barton Springs salamander (*Eurycea sosorum*), suggesting that planarians may be a significant but ignored prey item for aquatic salamander species/larvae [68].

The abundance of the crustaceans of the genus *Monolistra* was not only negatively related to fire salamander larvae occurrence, but also to the distance of the sites from the cave entrance. In particular we observed more *Monolistra* individuals in sites more distant from the surface. *Monolistra* is a genus of Sphaeromatidae that probably colonized caves form marine habitats [69]; it is possible that its occurrence is linked to older and stable aquifers and less linked to small groundwater sites close to the surface. Moreover, we also detected also a positive significant relationship between *Monolistra* abundance and the area of the pools. This variable might reflect the sampling effort as well as the surveyed suitable habitat. In some sites we observed high abundances of *Monolistra* that can constitute an important portion of the invertebrate biomass of groundwater. Further researches on these stygobionts are needed to understand their patterns of subterranean habitats exploitation.

The higher predation occurring close to surface may be one of the factors that limit the exploitation of interface habitats by groundwaters dwelling species when underground conditions are similar to those occurring in surface (e.g., during night or in intermediate seasons). However, the negative relationship observed between salamander larvae and stygofauna abundance could also be caused by non-consumptive effects of fire salamander and the landscape of fear generated by its occurrence. Moreover, analyses on interspecific/intraspecific interactions between stygobionts are required to understand how multispecies dynamics affect the abundance of the different invertebrate species. Finally, comparisons between the abundances in open pools/stream sectors and substrate/rocks interstices could provide further insights on the role played by microhabitat heterogeneity.

5. Conclusions

Salamander larvae can be a major predator for cave-adapted animals, with a keystone role at least in subterranean areas closer to the surface. However, these environments can be heavily impacted by ongoing climate changes, such as temperature increase and reduction of water availability, that may promote an increase in the use of caves by surface animals [70]. An increasing exploitation of caves by salamanders can have consequences on cave-adapted animals; thus understanding the role played by salamanders as predators can be central for the management of subterranean biodiversity at a broad scale. When in caves salamanders occupy the top predator level, thus they may have a top-down effect on other organisms not considered in this study. Future comparisons between the biofilms occurring in subterranean sites with and without fire salamander larvae could provide further insights on their cascading effects on cave trophic web.

Supplementary Materials: The following are available online at http://www.mdpi.com/1424-2818/12/4/161/s1, Table S1: Dataset fire salamander – stygofauna.

Author Contributions: Conceptualization, R.M.; data collection, R.M., B.B., A.M., M.F., E.L. and G.F.F.; data analysis, R.M and B.B.; writing—original draft preparation, R.M.; writing—review and editing E.L., M.F., G.F.F. and A.M. All authors have read and agreed to the published version of the manuscript.

Funding: This research was funded by THE MOHAMED BIN ZAYED SPECIES CONSERVATION FUND, grants numbers: 162514520; 180520056.

Acknowledgments: We are grateful to S. Salvidio for the invitation to contribute to this special issue. We thank E. Pezzoli, P. Pozzoli and the "Comitato Bevere" ong for help during surveys. E. Lunghi is supported by the Chinese Academy of Sciences President's International Fellowship Initiative for postdoctoral researchers (2019PB0143). The comments of S. Salvidio and of three anonymous reviewers improved the original version of the manuscript.

Conflicts of Interest: The authors declare no conflict of interest. The funders had no role in the design of the study; in the collection, analyses, or interpretation of data; in the writing of the manuscript, or in the decision to publish the results.

References

1. Wells, K.D. *The Ecology and Behaviour of Amphibians*; The University of Chicago Press: Chicago, IL, USA, 2007.
2. Davic, R.D.; Welsh, H.H. On the ecological roles of salamanders. *Annu. Rev. Ecol. Evol. Syst.* **2004**, *35*, 405–434. [CrossRef]
3. Wilbur, H.M. Experimental ecology of food webs: Complex systems in temporary ponds-The Robert H. MacArthur Award Lecture-Presented 31 July 1995 Snowbird. *Utah. Ecol.* **1997**, *78*, 2279–2302.
4. Anthony, C.; Hickerson, C.M.; Walton, B.M. Eastern Red-backed salamanders regulate top-down effects in a temperate forest-floor community. *Herpetologica* **2017**, *73*, 180–189.

5. Mancinelli, G.; Costantini, M.L.; Rossi, L. Top-down control of reed detritus processing in a lake littoral zone: Experimental evidence of a seasonal compensation between fish and invertebrate predation. *Int. Rev. Hydrobiol.* **2007**, *92*, 117–134. [CrossRef]

6. Petranka, J.W. *Salamanders of the United States and Canada. Washington*; Smithsonian Instute Press: Washington, DC, USA, 1998.

7. Crowther, T.W.; Stanton, D.W.; Thomas, S.M.; A'Bear, A.D.; Hiscox, J.; Jones, T.H.; Voriskova, J.; Baldrian, P.; Boddy, L. Top-down control of soil fungal community composition by a globally distributed keystone consumer. *Ecology* **2013**, *94*, 2518–2528. [CrossRef]

8. Walker, D.M.; Murray, C.M.; Talbert, D.; Tinker, P.; Graham, S.P.; Crowther, T.W. A salamander's top down effect on fungal communities in a detritivore ecosystem. *FEMS Microbiol. Ecol.* **2018**, *94*, fiy168. [CrossRef]

9. Fuge, R.; Perkins, W. Aluminium and heavy metals in potable waters of the north Ceredigion area, mid-Wales. *Environ. Geochem. Health* **1991**, *13*, 56–65. [CrossRef]

10. Geldreich, E.E. Drinking water microbiology–new directions toward water quality enhancement. *Int. J. Food Microbiol.* **1989**, *9*, 295–312. [CrossRef]

11. Culver, D.C.; Pipan, T. *The Biology of Caves and Other Subterranean Habitats*, 2nd ed.; Oxford University Press: New York, NY, USA, 2019.

12. Romero, A. *Cave Biology*; Cambridge University Press: Cambridge, UK, 2009.

13. Botello, A.; Iliffe, T.M.; Alvarez, F.; Juan, C.; Pons, J.; Jaume, D. Historical biogeography and phylogeny of Typhlatya cave shrimps (Decapoda: Atyidae) based on mitochondrial and nuclear data. *J. Biogeogr.* **2013**, *40*, 594–607. [CrossRef]

14. Trontelj, P.; Blejec, A.; Fiser, C. Ecomorphological Convergence of Cave Communities. *Evolution* **2012**, *66*, 3852–3865. [CrossRef]

15. Howarth, F.G.; Moldovan, O.T. The ecological classification of cave animals and their adaptations. In *Cave Ecology*; Moldovan, O.T., Kováč, L., Halse, S., Eds.; Springer: Berlin, Germany, 2018; pp. 41–67.

16. Gorički, S.; Niemiller, M.L.; Fenolio, D.B.; Gluesenkamp, A.G. Salamanders. In *Encyclopedia of Caves*; White, W.B., Culver, D.C., Pipan, T., Eds.; Academic Press: Cambridge, MA, USA, 2019; pp. 871–884.

17. Fenolio, D.B.; Graening, G.O.; Collier, B.A.; Stout, J.F. Coprophagy in a cave-adapted salamander; the importance of bat guano examined through nutritional and stable isotope analyses. *Proc. R. Soc. B Biol. Sci.* **2006**, *273*, 439–443. [CrossRef] [PubMed]

18. Ficetola, G.F.; Lunghi, E.; Canedoli, C.; Padoa-Schioppa, E.; Pennati, R.; Manenti, R. Differences between microhabitat and broad-scale patterns of niche evolution in terrestrial salamanders. *Sci. Rep.* **2018**, *8*, 10575. [CrossRef] [PubMed]

19. Lunghi, E.; Cianferoni, F.; Ceccolini, F.; Mulargia, M.; Cogoni, R.; Barzaghi, B.; Cornago, L.; Avitabile, D.; Veith, M.; Manenti, R.; et al. Field-recorded data on the diet of six species of European Hydromantes cave salamanders. *Sci. Data* **2018**, *5*, 180083. [CrossRef]

20. Niemiller, M.L.; Osbourn, M.S.; Fenolio, D.B.; Pauley, T.K.; Miller, B.T.; Holsinger, J.R. Conservation Status and Habitat Use of the West Virginia Spring Salamander (Gyrinophilus Subterraneus) and Spring Salamander (G. Porphyriticus) in General Davis Cave, Greenbrier Co., West Virginia. *Herpetol. Conserv. Biol.* **2010**, *5*, 32–43.

21. Manenti, R.; Ficetola, G.F.; Marieni, A.; de Bernardi, F. Caves as breeding sites for *Salamandra salamandra*: Habitat selection, larval development and conservation issues. *N. West. J. Zool.* **2011**, *7*, 304–309.

22. Manenti, R.; Ficetola, G.F. Salamanders breeding in subterranean habitats: Local adaptations or behavioural plasticity? *J. Zool.* **2013**, *289*, 182–188. [CrossRef]

23. Lunghi, E.; Manenti, R.; Ficetola, G.F. Seasonal variation in microhabitat of salamanders: Environmental variation or shift of habitat selection? *PeerJ* **2015**, *3*, e1122. [CrossRef]

24. Salvidio, S.; Costa, A.; Oneto, F.; Pastorino, M.V. Variability of a subterranean prey-redator community in space and time. *Diversity* **2020**, *12*, 17. [CrossRef]

25. Babik, W.; Rafinski, J. Amphibian breeding site characteristics in the Western Carpathians, Poland. *Herpetol. J.* **2001**, *11*, 41–51.

26. Manenti, R.; Melotto, A.; Denoël, M.; Ficetola, G.F. Amphibians breeding in refuge habitats have larvae with stronger antipredator responses. *Anim. Behav.* **2016**, *118*, 115–121. [CrossRef]

27. Steinfartz, S.; Weitere, M.; Tautz, D. Tracing the first step to speciation: Ecological and genetic differentiation of a salamander population in a small forest. *Mol. Ecol.* **2007**, *16*, 4550–4561. [CrossRef] [PubMed]

28. Limongi, L.; Ficetola, G.F.; Romeo, G.; Manenti, R. Environmental factors determining growth of salamander larvae: A field study. *Curr. Zool.* **2015**, *61*, 421–427. [CrossRef]

29. Manenti, R.; Lunghi, E.; Ficetola, G.F. Cave exploitation by an usual epigean species: A review on the current knowledge on fire salamander breeding in cave. *Biogeographia* **2017**, *32*, 31–46. [CrossRef]

30. Manenti, R.; Pennati, R.; Ficetola, G.F. Role of density and resource competition in determining aggressive behaviour in salamanders. *J. Zool.* **2015**, *296*, 270–277. [CrossRef]

31. Manenti, R.; Siesa, M.E.; Ficetola, G.F. Odonata occurence in caves: Active or accidentals? A new case study. *J. Cave Karst Stud.* **2013**, *75*, 205–209. [CrossRef]

32. Barzaghi, B.; Ficetola, G.F.; Pennati, R.; Manenti, R. Biphasic predators provide biomass subsidies in small freshwater habitats: A case study of spring and cave pools. *Freshw. Biol.* **2017**, *62*, 1637–1644. [CrossRef]

33. Culver, D.C.; Pipan, T. *Shallow Subterranean Habitats Ecology, Evolution, and Conservation*; Oxford University Press: New York, NY, USA, 2014.

34. Manenti, R.; Pezzoli, E. Think of what lies below, not only of what is visible above, or: A comprehensive zoological study of invertebrate communities of spring habitats. *Eur. Zool. J.* **2019**, *86*, 272–279. [CrossRef]

35. Pezzoli, E. I Molluschi crenobionti e stigobionti presenti in Italia. Censimento delle stazioni: VII aggiornamento. *Quad. Della Civ. Stn. Idrobiol. Milano* **1996**, *21*, 111–118.

36. Lunghi, E.; Corti, C.; Mulargia, M.; Zhao, Y.; Manenti, R.; Ficetola, G.F.; Veith, M. Cave morphology, microclimate and abundance of five cave predators from the Monte Albo (Sardinia, Italy). *Biodivers. Data J.* **2020**, *8*, e48623. [CrossRef]

37. Manenti, R.; de Bernardi, F.; Ficetola, G.F. Pastures vs forests: Do traditional pastoral activities negatively affect biodiversity? The case of amphibians communities. *N. West. J. Zool.* **2013**, *9*, 284–292.

38. Arcangeli, A. Note su alcuni sferomidi cavernicoli italiani. Bollettino dei Musei di zoologia e anatomia comparata della R. *Univ. Di Torino* **1942**, *49*, 117–125.

39. Stoch, F. Isopodi ed anfipodi (Crustacea, Malacostraca) della Provincia di Bergamo: Note sulle specie rinvenute nelle grotte e nelle sorgenti. In *I Molluschi Delle Sorgenti e Delle 'Acque Sotterranee', IX Aggiornamento al Censimento*; Pezzoli, E., Spelta, F., Eds.; Monografie di Natura Bresciana: Brescia, Italy, 2000; pp. 231–241.

40. Luštrik, R.; Turjakl, M.; Kralj-Fišer, S.; Fišer, C. Coexistence of surface and cave amphipods in an ecotone environment. *Contrib. Zool.* **2011**, *80*, 133–141. [CrossRef]

41. Fišer, C.; Kovačec, Ž.; Pustovrh, M.; Trontelj, P. The role of predation in the diet of *Niphargus* (Amphipoda: Niphargidae). *Speleobiol. Notes* **2010**, *2*, 4–6.

42. Manenti, R.; Barzaghi, B.; Lana, E.; Stocchino, G.A.; Manconi, R.; Lunghi, E. The stenoendemic cave-dwelling planarians (Platyhelminthes, Tricladida) of the Italian Alps and Apennines: Conservation issues. *J. Nat. Conserv.* **2018**, *45*, 90–97. [CrossRef]

43. Barker, R.J.; Schofield, M.R.; Link, W.A.; Sauer, J.R. On the reliability of N-Mixture models for count data. *Biometrics* **2017**, *74*, 369–377. [CrossRef]

44. Kuznetsova, A.; Brockhoff, P.B.; Christensen, R.H.B. lmerTest Package: Tests in Linear Mixed Effects Models. *J. Stat. Softw.* **2017**, *82*, 1–26. [CrossRef]

45. Brooks, M.E.; Kristensen, K.; van Benthem, K.J.; Magnusson, A.; Berg, C.W.; Nielsenn, A.; Skaug, H.J.; Maechler, M.; Bolker, B. glmmTMB Balances Speed and Flexibility Among Packages for Zero-inflatedn Generalized Linear Mixed Modeling. *R J.* **2017**, *9*, 378–400. [CrossRef]

46. Fox, J.; Weisberg, S. *An {R} Companion to Applied Regression*, 3rd ed.; Sage: Thousand Oaks, CA, USA, 2019.

47. Manenti, R.; Denoël, M.; Ficetola, G.F. Foraging plasticity favours adaptation to new habitats in fire salamanders. *Anim. Behav.* **2013**, *86*, 375–382. [CrossRef]

48. Manenti, R.; Ficetola, G.F.; Bianchi, B.; de Bernardi, F. Habitat features and distribution of *Salamandra salamandra* in underground springs. *Acta Herpetol.* **2009**, *4*, 143–151.

49. Melotto, A.; Ficetola, G.F.; Manenti, R. Safe as a cave? Intraspecific aggressiveness rises in predator-devoid and resource-depleted environments. *Behav. Ecol. Sociobiol.* **2019**, *73*, 68. [CrossRef]

50. Costa, A.; Baroni, D.; Romeno, A.; Salvidio, S. Individual diet variation in *Salamandra salamandra* (Linnaeus, 1758) larvae in a Mediterranean stream. *Salamandra* **2017**, *53*, 148–152.

51. Reinhardt, T.; Steinfartz, S.; Paetzold, A.; Weitere, M. Linking the evolution of habitat choice to ecosystem functioning: Direct and indirect effects of pond-reproducing fire salamanders on aquatic-terrestrial subsidies. *Oecologia* **2013**, *173*, 281–291. [CrossRef] [PubMed]

52. Costa, A.; Salvidio, S.; Romano, A.; Baroni, D. Larval diet of *Salamandra salamandra* (L., 1758): Preliminary results on prey selection and feeding strategy. In Proceedings of the Atti X congresso Nazionale della Societas Herpetologica Italica, Genova, Italy, 15–18 October 2014; Doria, G., Poggi, R., Salvidio, S., Tavano, M., Eds.; Ianieri Edizioni: Pescara, Italy, 2014; pp. 33–38.

53. Mosslacher, F. Subsurface dwelling crustaceans as indicators of hydrological conditions, oxygen concentrations, and sediment structure in an alluvial aquifer. *Int. Rev. Hydrobiol.* **1998**, *83*, 349–364. [CrossRef]

54. Durkota, J.M.; Wood, P.J.; Johns, T.; Thompson, J.R.; Flower, R.J. Distribution of macroinvertebrate communities across surface and groundwater habitats in response to hydrological variability. *Fundam. Appl. Limnol.* **2019**, *193*, 79–92. [CrossRef]

55. Galassi, D.M.P.; Stoch, F.; Fiasca, B.; di Lorenzo, T.; Gattone, E. Groundwater biodiversity patterns in the Lessinian Massif of northern Italy. *Freshw. Biol.* **2009**, *54*, 830–847. [CrossRef]

56. Manenti, R.; Lunghi, E.; Ficetola, G.F. Distribution of spiders in cave twilight zone depends on microclimatic features and trophic supply. *Invertebr. Biol.* **2015**, *134*, 242–251. [CrossRef]

57. Lunghi, E.; Manenti, R.; Ficetola, G.F. Cave features, seasonality and subterranean distribution of non-obligate cave dwellers. *PeerJ* **2017**, *5*, e3169. [CrossRef]

58. Salvidio, S.; Palumbi, G.; Romano, A.; Costa, A. Safe caves and dangerous forests? Predation risk may contribute to salamander colonization of subterranean habitats. *Sci. Nat.* **2017**, *104*, 20. [CrossRef]

59. Fišer, C. *Niphargus*—A model system for evolution and ecology. In *Encyclopedia of Caves*; White, W.B., Culver, D.C., Pipan, T., Eds.; Academic Press: Cambridge, MA, USA, 2019; pp. 746–755.

60. Väinölä, R.; Witt, J.D.S.; Grabowski, M.; Bradbury, J.H.; Jażdżewski, K.; Sket, B. Global diversity of amphipods (Amphipoda; Crustacea) in freshwater. *Hydrobiologia* **2008**, *595*, 241–255. [CrossRef]

61. Marković, V.; Novaković, B.; Ilić, M.; Nikolić, V. Epigean Niphargids in Serbia: New Records of *Niphargus valachicus* Dobreanu & Manolache, 1933 (Amphipoda: Niphargidae), with Notes on its Ecological Preferences. *Acta Zool. Bulg.* **2018**, *70*, 45–50.

62. Fišer, C.; Keber, R.; Kerezi, V.; Moskric, A.; Palandancic, A.; Petkovska, V.; Potocnik, H.; Sket, B. Coexistence of species of two amphipod genera: *Niphargus timavi* (Niphargidae) and *Gammarus fossarum* (Gammaridae). *J. Nat. Hist.* **2007**, *41*, 2641–2651. [CrossRef]

63. Fišer, Z.; Novak, L.; Lustrik, R.; Fiser, C. Light triggers habitat choice of eyeless subterranean but not of eyed surface amphipods. *Sci. Nat.* **2016**, *103*, 7. [CrossRef]

64. Romeo, G.; Giovine, G.; Ficetola, G.F.; Manenti, R. Development of the fire salamander larvae at the altitudinal limit in Lombardy (north-western Italy): Effect of two cohorts occurrence on intraspecific aggression. *N. West. J. Zool.* **2015**, *11*, 234–240.

65. Steinfartz, S.; Stemshorn, K.; Kuesters, D.; Tautz, D. Patterns of multiple paternity within and between annual reproduction cycles of the fire salamander (*Salamandra salamandra*) under natural conditions. *J. Zool.* **2006**, *268*, 1–8. [CrossRef]

66. Manenti, R.; Barzaghi, B.; Tonni, G.; Ficetola, G.F.; Melotto, A. Even worms matter: Cave habitat restoration for a planarian species has increased prey availability but not population density. *Oryx* **2019**, *53*, 216–221. [CrossRef]

67. Reynoldson, J.D.; Young, J.O. *A key to the Freshwater Triclads of Britain and Ireland with Notes on Their Ecology*; Freshwater Biological Association: Ambleside, UK, 2000.

68. Gillespie, J.H. Application of stable isotope analysis to study temporal changes in foraging ecology in a highly endangered amphibian. *PLoS ONE* **2013**, *8*, 10. [CrossRef]

69. Prevorčnik, S.; Verovnik, R.; Zagmajster, M.; Sket, B. Biogeography and phylogenetic relations within the Dinaric subgenus Monolistra (Microlistra) (Crustacea: Isopoda: Sphaeromatidae), with a description of two new species. *Zool. J. Linn. Soc.* **2010**, *159*, 1–21. [CrossRef]

70. Mammola, S.; Piano, E.; Cardoso, P.; Vernon, P.; Dominguez, D.; Isaia, M. Climate change going deep: The effects of global climatic alterations on cave ecosystems. *Anthr. Rev.* **2019**, *6*, 98–116. [CrossRef]

 *diversity*

Article

Variability of A Subterranean Prey-Predator Community in Space and Time

Sebastiano Salvidio [1,2,*], **Andrea Costa** [1], **Fabrizio Oneto** [1,2] **and Mauro Valerio Pastorino** [2]

[1] Dipartimento Scienze della Terra dell'Ambiente e della Vita—DISTAV, Università degli Studi di Genova, 16132 Genova, Italy; andrea-costa-@hotmail.it (A.C.); oneto.fabrizio@alice.it (F.O.)

[2] Gruppo Speleologico "A. Issel", Villa Comunale ex Borsino c.p. 21, 16012 BUSALLA (GE), Italy; mvpastor@tiscali.it

* Correspondence: sebastiano.salvidio@unige.it; Tel.: +39-010-3358027

Received: 18 December 2019; Accepted: 27 December 2019; Published: 31 December 2019

Abstract: Subterranean habitats are characterized by buffered climatic conditions in comparison to contiguous surface environments and, in general, subterranean biological communities are considered to be relatively constant. However, although several studies have described the seasonal variation of subterranean communities, few analyzed their variability over successive years. The present research was conducted inside an artificial cave during seven successive summers, from 2013 to 2019. The parietal faunal community was sampled at regular intervals from outside to 21 m deep inside the cave. The community top predator is the cave salamander *Speleomantes strinatii*, while invertebrates, mainly adult flies, make up the rest of the faunal assemblage. Our findings indicate that the taxonomic composition and the spatial distribution of this community remained relatively constant over the seven-year study period, supporting previous findings. However, different environmental factors were shaping the distribution of predators and prey along the cave. Invertebrates were mainly affected by the illuminance, while salamanders were influenced by both illuminance and distance from the cave's entrance. The inter-annual spatial distribution of the salamander population was highly repeatable and age specific, confirming a gradual shift towards the deeper parts of the cave with an increasing age. In general, the spatial distribution along the cave of this prey-predator system remained relatively constant during the seven-year study, suggesting that strong selective constraints were in action, even in this relatively recent subterranean ecosystem.

Keywords: artificial cave; ecotone; prey-predator system; salamanders; *Speleomantes*; subterranean habitat

1. Introduction

Subterranean habitats are simplified ecosystems characterized by reduced climatic fluctuations in comparison to those occurring in the surrounding surface habitats [1–3]. In subterranean habitats, air temperature and relative humidity display buffered seasonal variations. Solar radiation is completely absent in deep areas, yet still present but dimmed in the twilight ecotone zone [4–6]. In addition, subterranean environments are often energy limited in the sense that, in absence of primary producers, the main organic supply is derived from organisms living in surrounding surface habitats [1]. Therefore, the organic basis of the food chain in subterranean ecosystems is provided by active movements of animals that periodically migrate inside the system during external unfavorable periods or is imported as organic debris by gravity, wind, and rainflow from outside [7,8]. Several studies analyzed the influence of seasonal climatic variations on the abundance and distribution of single or few subterranean populations or species (e.g., references [4,9–12] or, more in general, on the composition of entire biological communities living in subterranean environments e.g., references [5,13–15]. Concerning the long-term composition of biological cave communities, it is generally assumed that those living in subterranean habitats are characterized by low levels of temporal variability [1,16]. This assumption was analyzed in detail by Di Russo et al. [16], who studied the temporal variations of the faunal community living inside the natural cave of Valdemino in central Italy. Di Russo et al. [16] found that the resident cave community, composed exclusively by invertebrates, maintained a relatively constant species composition and displayed a constant spatial distribution over the 20-year study period. However, there are some exceptions to the assumption of low temporal variability of subterranean communities. One convincing example is the cricket guano community studied inside Mammoth Cave in Kentucky (USA) by [17]. This study described the observed changes in species composition over more than two decades. Changes were gradual and apparently not influenced by local factors, such as human disturbance, but apparently were caused by directional effects of unfavorable climatic events on the population of the cave cricket *Hadenoecus subterraneus*, the key species providing organic supply to the entire subterranean community [17]. Therefore, although the general hypothesis of low temporal variability of subterranean communities is well accepted [2], this hypothesis should be verified on a case-by-case basis. This because, when studying the composition and distribution of subterranean communities, the great heterogeneity in the physical structure and local climate of these habitat has to be taken into consideration. In fact, subterranean habitats are very heterogeneous and may vary in the size, number, and orientation of their connections with the surface, vertical or horizontal development of penetrable spaces, local climatic factors, extension of the twilight zone, and availability of trophic subsidies [8,18,19].

In this paper, we describe variations of the composition and spatial distribution of a biological community sampled inside an artificial cave. The faunal community is composed by a terrestrial top predator, the salamander *Speleomantes strinatii* (Aellen, 1958), and its invertebrate prey taxa [20]. This community was sampled each year in July with the same methodology to allow a robust among-year comparison. In fact, artificial subterranean habitats constitute interesting ecosystems, because they are relatively young and their age is usually well known. Moreover, these artificial habitats are generally characterized by environmental conditions similar to those of natural caves found in the same geographic area, in particular when the variation of air relative humidity, air temperature and direct solar radiation are considered [21].

The aim of this study was two-fold: (i) to assess the changes in the biological community over seven consecutive summers and to validate the hypothesis of low inter-annual variability of biological communities, and (ii) to evaluate which environmental factors were shaping the different ecological groups composing the community and whether the spatial distribution of the top predator, the cave salamander, and of its invertebrate prey was influenced by similar or contrasting environmental factors.

2. Materials and Methods

The study site is situated at 369 m a.s.l. in the municipality of Savignone (Province of Genova, Region of Liguria, Italy). This artificial cave is a U-shaped tunnel that develops horizontally for about 40 m. The site was excavated in a geological substratum composed by thin layers of siltstone and claystone, attributed to a late Cretaceous period known as Campanian [22]. This cave was excavated to be used as an air-raid shelter during World War II (i.e., in the period 1941–1943) and originally had two large entrances [20,23], but one of them collapsed soon after the shelter construction. Since 1987, the only cave entrance left was closed by an iron gate and the tunnel's walls were equipped with a permanent grid with a 1 × 1 m mesh, to allow studying the salamander population that lives in the cave [24]. This underground laboratory is managed by the Speleological Group "A. Issel" and is named "Biospeleological Station of San Bartolomeo". The seasonal temperature variations recorded inside the cave display similar patterns in comparison to those occurring in natural caves with only a single large entrance [21,25]. In this particular case, air temperatures measured over 12 consecutive months in different parts of the study cave showed a similar, but much more buffered pattern of variation, in comparison to those recorded outside at the nearest meteorological station (Figure 1). In fact, the coefficient of variation (CV) of the air temperature at the cave entrance was already halved (CV = 0.29) and more than fourfold lower at 16 m deep inside (CV = 0.15), when compared with the corresponding temperature variation (CV = 0.67) measured at the surface [23] (Table 1).

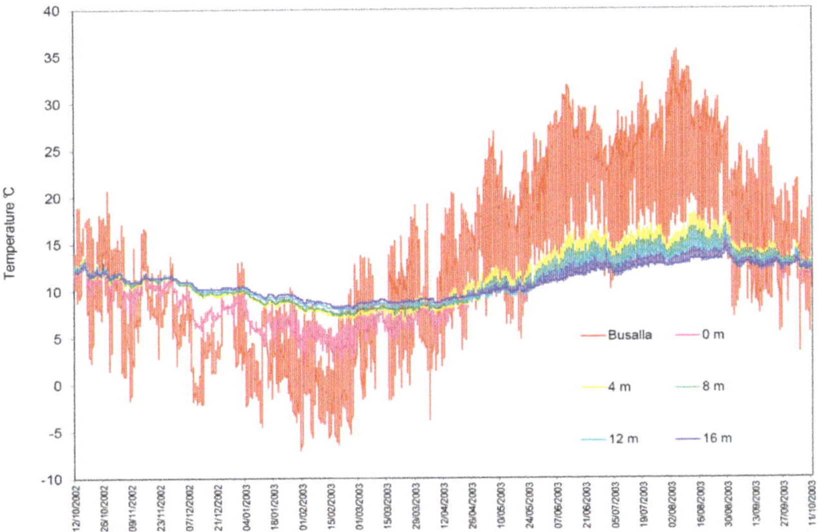

Figure 1. Air temperatures recorded inside and outside the "Biospeleological Station of San Bartolomeo". Busalla refers to the meteo-station of Busalla (Province of Genova). The recording period was October 12, 2002 – October 11, 2003. The data-loggers used were Hanna Instruments HI 140. 0 m = cave entrance; 4, 8, 12, and 16 are meters from the cave entrance.

Table 1. Number of temperature records, minimum, maximum, mean, standard deviation and coefficient of variation of the annual temperatures recorded outside and inside the study site of the "Biospeleological Station of San Bartolomeo", from October 2002 to October 2003. Table modified from reference [23].

	Busalla Meteo-Station	Biospelological Station of Besolagno				
		0 m from entrance	4 m from entrance	8 m from entrance	12 m from entrance	16 m from entrance
T record number	7306	7600	7600	7600	7600	7600
T minimum (°C)	−7.0	2.1	6.6	7.3	7.8	8.1
T maximum (°C)	35.6	15.3	18.2	17.0	16.7	15.3
T mean (°C)	12.6	10.0	11.3	11.2	11.1	10.9
Standard deviation	8.9	2.9	2.7	2.4	2.1	1.7
Coefficient of variation	68.4	29.4	24.3	21.6	18.9	15.1

In this study, three environmental parameters characterizing the different parts of the cave were quantified. The variation in solar radiation along the cave was measured in Lux (LUX) with a Unittic photometer at midday in July 2015. The linear distance from entrance (DISTANCE) was recorded in meters through a permanent grid with a 1 × 1 m mesh as reported by reference [20]. The complexity of cave walls (COMPLEXITY) was expressed in cm, by summing the linear distance attained by a 1 m string pressed vertically on the left and right cave walls at about 1.5 m of height [4,15]. These measures were then subtracted from 200, the maximum value obtained if both cave walls were completely smooth (i.e., low complexity). In this way a low substrate complexity will score 0, while increasing complexity gradually will score increasing larger values (Supplementary Materials Table S1).

The study of the salamander population abundance and distribution is part of a long-term research that began in 1996 and is still ongoing [26,27]. The data used in this study are those obtained from 2013 to 2019. Absolute population abundance was estimated, every year in July, by a three-occasion temporary removal experiment in which samples were obtained every other day [26,28]. Salamanders were caught by hand on the cave walls, measured to the nearest millimeter (snout-vent length, SVL) and caged inside terraria until the end of the sampling. All animals were returned unharmed to their cave section at the end of the annual removal experiment. Abundance was estimated using model M_{bh} in CAPTURE software [29]. Individuals possessing a swollen mental gland were identified as adult males [30], while individuals larger than 57 mm in SVL were considered adult females [31]. Each year, the SVL polymodal distribution of the immature portion of the population was decomposed by the use of FiSAT software [32]. Juveniles aged one or two (thereafter "juveniles") were pooled and separated from sub-adults (i.e., large salamanders in their third year, but not yet sexually mature) by taking into account yearly cohort distributions [28]. The relative abundance of juveniles, subadults, females, and males that were caught by hand in the cave section corresponding to each adhesive entomological trap (see below) was obtained by summing the individuals removed over the three sampling occasions. These numbers represented almost a complete census, because during this study, salamander capture probabilities were relatively high, being 0.54 on average (see Results and Supplementary Materials Table S2).

Each year, during the first day of the salamander removal, nine adhesive traps were positioned outside, at the entrance and at 3, 6, 9, 12, 15, 18, and 21 m inside the cave. Each trap was a transparent acetate sheet (21 × 30 cm) coated on one side with entomological glue, hanging from the cave vault. At the end of the salamander removal, all traps were retrieved and permanently conserved between transparent plastic sheets. Subsequently, trapped invertebrates were identified and counted in the laboratory under a dissecting microscope. The use of adhesive trap is possibly selective towards flying insects as dipterans. However, this technique was used because salamanders living in the study cave feed prevalently upon this type of prey, that constitutes about 70% of the total prey items ingested [20].

The variations of the entire biological community were analyzed by a two-way permutation multivariate analysis of variance (PERMANOVA) based on square root transformed data, Bray-Curtis dissimilarity, and 9999 permutations, by using PAST software [33]. To obtain a balanced design the traps located outside, at the entrance and at 3 m inside the cave were considered replicates of the light/twilight zone (i.e., where light is present or reduced), those at 6, 9, and 12 m were considered replicates of the dark zone (i.e., where light is absent), while traps at 15, 18, and 21 m were considered replicates of the deep zone (i.e., where light is absent and climatic conditions become constant). Thus, "year" and "cave zone" were used as factors, while abundances of invertebrate taxa and of the four salamander groups (i.e., juveniles, subadults, females, and males, Table 2) were the dependent variables.

The relationship between the faunal community and the three environmental variables was assessed by canonical correspondence analysis (CCA), a multivariate analysis specifically designed for the investigation of ecological gradients [34,35]. In particular, CCA selects the linear combinations of environmental variables that maximizes the species' dispersion and elaborates an ordination diagram, delimited by uncorrelated linear axes, in which species, sites and environmental variables are projected [5,34]. The overall robustness of CCA was calculated by 9999 permutations [35]. In our CCA, the biological community consisted in the four groups of salamanders, while the invertebrates where divided in two categories: dipterans (DI) and all the other taxa (OT) pooled. This procedure was justified because dipterans were numerically predominant in the study site [20] (see also Results).

Table 2. Faunal community sampled outside, at the entrance and at different distances inside the "Biospeleological Station of San Bartolomeo". The category "Undetermined insects" was not used in multivariate analyses.

	Outside	Entrance	3 m	6 m	9 m	12 m	15 m	18 m	21 m
Year 2013									
Diptera	62	5	1	0	0	2	0	1	1
Acarina	1	0	0	0	0	0	0	0	0
Hymenoptera	1	0	0	0	0	0	0	0	0
Coleoptera	7	0	0	0	0	0	0	0	0
Homoptera	9	0	0	0	0	0	0	0	0
Araneida	1	0	0	0	0	0	0	0	0
Undetermined insects	5	1	1	1	1	0	0	0	0
Salamander males	0	0	0	0	0	0	0	0	2
Salamander females	0	0	0	0	0	1	0	1	1
Salamander subadults	0	0	0	0	1	1	1	0	0
Salamander juveniles	0	0	5	2	2	1	0	0	0
Year 2014									
Diptera	31	3	0	2	0	1	0	0	1
Acarina	0	1	0	0	0	0	0	0	0
Hymenoptera	3	0	0	0	0	0	0	0	0
Coleoptera	3	0	0	0	0	0	0	0	0
Homoptera	3	0	0	0	0	0	0	0	0
Araneida	0	0	0	0	0	0	0	0	0
Undetermined insects	1	0	1	2	0	0	0	0	0
Salamander males	0	0	0	1	0	4	1	0	0
Salamander females	0	0	0	0	0	1	1	0	0
Salamander subadults	0	0	1	1	1	0	0	0	0
Salamander juveniles	0	0	7	12	0	1	0	0	0
Year 2015									
Diptera	30	1	2	3	0	2	3	5	2
Acarina	0	0	0	0	0	0	0	0	0
Hymenoptera	1	0	0	0	0	0	0	0	0
Coleoptera	4	0	0	0	0	0	0	0	0
Homoptera	1	0	0	0	0	0	0	0	0
Araneida	2	0	0	0	0	0	0	0	0
Undetermined insects	2	0	0	1	0	0	0	0	0
Salamander males	0	0	1	2	2	1	0	2	1
Salamander females	0	0	0	1	0	0	0	0	0
Salamander subadults	0	0	0	2	4	1	0	1	0
Salamander juveniles	0	0	6	2	0	1	0	0	0
Year 2016									
Diptera	41	10	2	2	4	3	8	10	9
Acarina	0	0	1	0	0	0	0	0	0
Hymenoptera	1	0	0	0	0	0	0	0	0
Coleoptera	6	0	0	0	0	0	0	0	0
Homoptera	3	0	0	0	0	0	0	0	0
Araneida	0	1	0	0	0	0	0	0	0
Undetermined insects	0	0	1	0	0	0	0	0	0
Salamander males	0	0	0	1	0	1	1	0	0
Salamander females	0	0	0	0	0	0	1	0	0
Salamander subadults	0	0	0	1	0	1	0	0	0
Salamander juveniles	0	0	4	4	0	1	0	0	0

Table 2. *Cont.*

	Outside	Entrance	3 m	6 m	9 m	12 m	15 m	18 m	21 m
Year 2017									
Diptera	64	2	0	0	1	1	1	1	1
Acarina	0	0	0	0	0	0	0	0	0
Hymenoptera	1	0	0	0	0	0	0	0	0
Coleoptera	8	0	0	0	0	0	0	0	0
Homoptera	32	2	0	0	0	0	0	0	0
Araneida	1	0	0	0	0	0	0	0	0
Undetermined insects	1	0	0	0	1	0	0	0	0
Salamander males	0	0	0	0	1	0	1	2	1
Salamander females	0	0	0	1	0	0	0	0	1
Salamander subadults	0	0	0	1	1	0	0	0	0
Salamander juveniles	0	0	4	3	0	0	0	0	0
Year 2018									
Diptera	40	3	7	2	2	20	21	25	44
Acarina	0	0	0	0	0	0	0	0	0
Hymenoptera	0	0	0	0	0	0	0	0	0
Coleoptera	1	0	0	0	0	0	0	0	0
Homoptera	7	0	1	0	0	0	0	0	0
Araneida	3	0	1	0	0	0	0	0	0
Undetermined insects	0	0	1	0	0	0	0	0	0
Salamander males	0	0	0	3	1	1	1	0	0
Salamander females	0	0	0	0	0	0	0	1	0
Salamander subadults	0	0	0	3	0	0	0	0	0
Salamander juveniles	0	0	1	4	2	0	0	0	0
Year 2019									
Diptera	11	1	0	1	3	3	4	3	0
Acarina	0	0	0	0	0	0	0	0	0
Hymenoptera	2	0	0	0	0	0	0	0	0
Coleoptera	3	0	0	0	0	0	1	0	0
Homoptera	9	1	0	0	0	0	0	0	0
Araneida	0	0	1	0	0	0	0	1	0
Undetermined insects	1	0	0	1	0	0	1	0	1
Salamander males	0	0	0	0	1	0	1	2	1
Salamander females	0	0	0	1	0	0	0	0	1
Salamander subadults	0	0	0	1	1	0	0	0	0
Salamander juveniles	0	0	4	3	0	0	0	0	0

3. Results

During the study period, the total estimated salamander abundance fluctuated from 98 in 2019 to 126 in both 2013 and 2014 (CV = 0.11). Capture probabilities were rather high, being 0.54 on average (95% confidence interval 0.49–0.60), a value indicating that after three removal samples, about 90% of the entire salamander population was captured and measured each year (Supplementary Materials, Table S2).

A total of 656 invertebrates were caught during this study, with 61% of them trapped outside the cave (Table 2). This trapping location was also characterized by the constant presence of sup-sucking aphids (Homoptera), a group of insects caught inside the cave only when passively transported by wind (Table 2). Flies (Diptera) were the most abundant taxon (N = 508), representing 77% of the total invertebrates and about 65% of invertebrates trapped inside the cave, many of which belonged to the common crane fly *Limonia nubeculosa*.

The results of the two-way PERMANOVA are reported in Table 3. The interaction between the trapping zone and the sampling year, and the biological community in different years were non-significant (Cave zone * Year: F = -0.65, p = 0.93; Year: F = 0.91, p = 0.13). Conversely, as expected, the faunal composition among the three sampling zones showed a highly significant degree of dissimilarity, confirming that there was a differentiation in the faunal composition along the subterranean habitat (F = 4.73, p = 0.0001). Overall, these findings suggested that there were no relevant variations in the annual composition of the faunal assemblage in the study site, and that the community spatial distribution remained relatively constant along the cave during the seven-year study period.

Table 3. Results of the two-way PERMANOVA (9999 permutations) comparing the temporal and spatial variation of the faunal community sampled along the Biospeleological Station of San Bartolomeo from 2013 to 2019.

	Sum of Squares	df	Mean Square	F	p
Year	1.7389	6	0.2898	0.9060	0.1336
Cave zone	3.0232	2	1.5116	4.7254	0.0001
Year * Cave zone	−2.4987	12	−0.2082	−0.6509	0.9338
Residual	13.115	41	0.3199		
Total	15.379	61			

The CCA using the environmental variables and the entire biological community sampled along the subterranean gradient was highly significant (trace = 0.765; p = 0.001; Figure 2). The first two axes explained the entire variance of the system (67% and 33%, respectively), and both were significant (axis 1: eigenvalue = 0.509, p = 0.008; axis 2: eigenvalue = 0.256, p = 0.003).

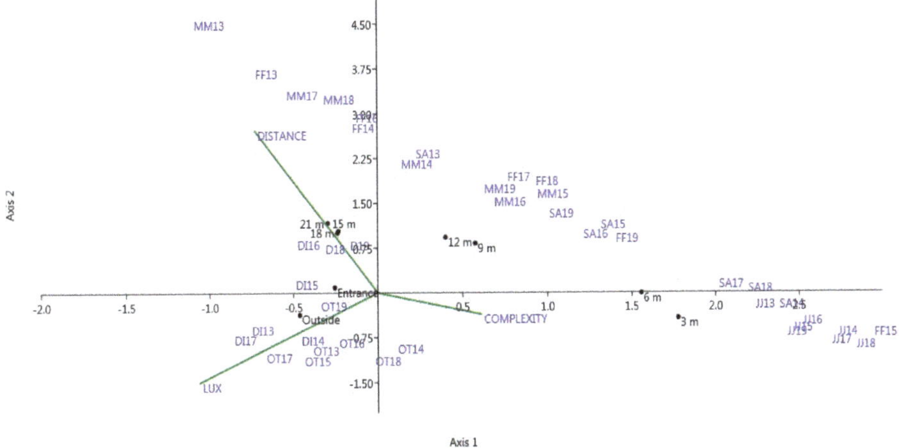

Figure 2. Canonical correspondence ordination diagram of the faunal community sampled from 2013 to 2019 in the Biospeleological Station of San Bartolomeo. DI = dipterans, FF = female salamanders, JJ = juvenile salamanders, MM = male salamanders, OT = other invertebrate taxa, SA = subadult salamanders. The number following the symbols indicate the year of sampling. COMPLEXITY is the wall heterogeneity, DISTANCE is the linear distance from the entrance and LUX is the illuminance.

Dipterans (DI) and the other taxa of invertebrates (OT) were projected in the left lower quadrant of the plot, while salamanders were linearly spread in along the diagram, with juveniles and subadults always projected on the right and females and males in the center and left quadrant (Figure 2). There was a strong repeatability of this pattern among years with only one exception, the position of female salamanders sampled in 2015 (i.e., FF15 in Figure 2), that was associated with juveniles and subadults on the rightmost part of the plot. The distribution of trapping sites on the CCA plot was also noteworthy, because the two more external sites (i.e., Outside and Entrance) were projected in the lower left quadrant in association with all the invertebrate taxa. Conversely, all the others sites were distributed on the right part and the upper part of the CCA plot, apparently associated with the salamanders' groups. Concerning environmental factors, LUX had the highest load on the first axis, while DISTANCE had the highest load on the second axis (Supplementary materials; Table S3). Finally, the complexity of the cave walls (COMPLEXITY) had apparently a trivial effect on the entire faunal community distribution, because this factor displayed its highest load on the third axis that was non-significant ($p = 0.9532$; Supplementary materials Table S3).

4. Discussion

According to Mammola and Isaia [36] the number of species found in subterranean habitats, their relative abundance and their ecological interactions are likely to depend on the carrying capacity of the system, i.e., on the amount of energy entering the system from outside. Our study habitat had a recent origin, being only 70 years old, and its physical structure was relatively simple, possessing only one large entrance and being horizontal. Furthermore, the entire ecological community was constituted by non-specialized (or "troglophile" *sensu* [37]) forest or soil animals that colonize opportunistically newly created subterranean habitats in search of new ecological opportunities, to avoid unfavorable climatic conditions, or to reduce competition and predation [38–40]. Our study showed a general constancy in both composition and distribution along the cave of the entire faunal community, at least during the seven-year period investigated. Therefore, the starting hypothesis of a low inter-annual variability characterizing subterranean biological communities [2,14,16] was corroborated by our multivariate approach. Consequently, our results seem to extend the hypothesis of low temporal variability of subterranean ecosystems to recently-established biological communities colonizing human-made habitats, in particular those characterized by highly buffered climatic conditions. In fact, the PERMANOVA clearly showed that there were no annual changes in the community composition within the three different zones in which the study cave was subdivided, i.e., the twilight, dark, and deep zones. This finding is of particular interest when considering that the transitional twilight zone analyzed in this study was in connection to the exterior environment through a large entrance and therefore, highly subject to abrupt and unpredictable seasonal and even daily climatic variations [6].

Another relevant finding concerns the different influences of the environmental factors on the prey-predator community found in the subterranean habitat. In particular, the spatial distribution of the top predator, the salamander *Speleomantes strinatii*, seemed influenced by contrasting environmental factors in comparison to its invertebrate prey taxa (see Figure 2). In fact, only illuminance affected the spatial distribution of invertebrates. Conversely, the distribution of cave salamanders was apparently caused by the interaction between illuminance and distance from the entrance. In the study site, these two factors are partially decoupled because light is completely and permanently lacking at a distance of 9 m from the entrance (Supplementary materials, Table S1). Thus, the deeper sections of the cave are homogeneous for this factor and can be differentiated mainly by a decreasing level of environmental variability (Table 1) and possibly by a different level of air humidity. Moreover, while illuminance had a clear positive effect on invertebrate distribution, its effect was opposite on salamanders, even for juveniles that were often abundant near the entrance, but were also found along the study cave in the dark zone up to 12 m were light is completely absent (Table 1). Thus, different ecological conditions were related to the spatial distribution of salamanders of all ages and their invertebrate prey. It is plausible that other environmental non-measured factors, such as air humidity [15,41] and the low climatic variability recorded in the deeper part of the cave, were affecting the observed distribution of salamanders. Other studies have analyzed the spatial distribution of cave salamanders within subterranean habitats in other parts of Italy [15,41,42]. In these studies, however, a different species of salamander (*Speleomantes italicus*) was studied and, in addition, a different sampling design was performed, as data from several different caves were pooled in statistical analyzes over only one study year. In the study of Lunghi et al. [15], cave salamanders were strictly associated with high relative air humidity and with distance from cave entrance. In another similar study, juvenile salamanders appeared to be strongly associated with invertebrate prey [41]. While our results are in part consistent with those of Lunghi et al. in some cases [15,41], with adult salamanders associated with deep and more stable climatic subterranean sectors, in other cases our findings suggest a somewhat different scenario. In fact, in our study site, invertebrates and salamanders were affected by different environmental factors, and there was no strict association between juvenile salamanders and invertebrates (Figure 2). These findings could be due to the different sampling scheme used, but also to the absence of dipterans in the invertebrate dataset of Lunghi et al. [15]. However, a common conclusion of all studies was that wall roughness had little influence on cave salamander distribution along the entire cave development. Apparently, climatic factors are much more relevant than the physical structure of the cave rock walls in shaping the underground distribution of Italian cave salamanders.

Concerning specifically the cave salamander *S. strinatii*, its population structure comprised all age groups, and in particular by a large proportion of juveniles (Supplementary materials, Table S2; [26,28]. This demographic structure was observed in all years (Supplementary materials, Table S2), indicating that the population has permanently colonized this artificial habitat and is successfully reproducing there.

In conclusion, the composition and spatial distribution of the entire faunal community living inside the study cave site appeared to be relatively constant over the seven-year study. As already observed by Romero [43,44] and recently reviewed by Mammola [45], subterranean habitats should be considered open rather than closed ecosystems, and they should be analyzed by taking into consideration the physical and biological features of adjacent surface habitats, where several subterranean populations migrate to forage during favorable periods and from where many organisms are constantly entering subterranean habitats to shelter from unfavorable conditions or to reproduce. In the present case, the study cave was recently built and subject to a continuous exchange of troglophile organisms from surrounding surface habitats. Notwithstanding this, a relatively constant biological community was observed, indicating that in this subterranean habitat, strong selective constraints were acting and were stabilizing this recently formed subterranean food web in both space and time.

Supplementary Materials: The following are available online at http://www.mdpi.com/1424-2818/12/1/17/s1: Table S1: Physical parameters measured outside and inside the experimental cave; Table S2: Removal statistics of the cave salamander *Speleomantes strinatii* population living in the "Biospeleological Station of San Bartolomeo"; Table S3: Scores of the Canonical Correspondence Analysis.

Author Contributions: Conceptualization, S.S., F.O. and M.V.P.; Methodology and sampling, S.S., F.O.; Formal analysis, S.S. and A.C.; Data curation, S.S.; Writing—original draft preparation, S.S. and A.C.; Writing—review and editing, S.S., A.C., F.O. and M.V.P. All authors have read and agreed to the published version of the manuscript.

Funding: This research received no external funding.

Acknowledgments: Permits to temporary capture and manipulate the salamanders were provided by the Italian Ministry of Environment (42466/PNM of 2013 and 00004038/PNM of 2019). We thank two anonymous Reviewers and Žiga Fišer for comments that improved a previous version of the manuscript.

Conflicts of Interest: The authors declare no conflict of interest.

References

1. Poulson, T.L.; White, W.B. The cave environment. *Science* **1969**, *165*, 971–981. [CrossRef]
2. Culver, D.C. *Cave Life*; Harvard University Press: Cambridge, MA, USA, 1982.
3. Smithson, P.A. Inter-relationships between cave and outside air temperatures. *Theor. Appl. Climatol.* **1991**, *44*, 65–73. [CrossRef]
4. Camp, C.D.; Jensen, J.B. Use of twilight zones of caves by plethodontid salamanders. *Copeia* **2007**, *2007*, 594–604. [CrossRef]
5. Tobin, B.W.; Hutchins, B.T.; Schwartz, B.F. Spatial and temporal changes in invertebrate assemblage structure from the entrance to deep-cave zone of a temperate marble cave. *Int. J. Speleol.* **2013**, *42*, 203–214. [CrossRef]
6. Mammola, S.; Isaia, M. Day–night and seasonal variations of a subterranean invertebrate community in the twilight zone. *Subterr. Biol.* **2018**, *27*, 31–51. [CrossRef]
7. Poulson, T.L. Food sources. In *Encyclopedia of Caves*; Culver, D.C., White, W.B., Eds.; Elsevier Academic Press: Burlington, MA, USA, 2007; pp. 255–264.

8. Coulver, D.C. Entrances. In *Encyclopedia of Caves*; Culver, D.C., White, W.B., Eds.; Elsevier Academic Press: Burlington, MA, USA, 2007; pp. 206–208.

9. Keith, J.H. Seasonal changes in a population of Pseudanophtalmus tenuis (Coleoptera, Carabidae) in Murray Spring Cave, Indiana: A preliminary report. *Int. J. Speleol.* **1975**, *7*, 33–44. [CrossRef]

10. Crouau-Roy, B.; Crouau, Y.; Ferre, C. Dynamic and temporal structure if the troglobitic beetle Speonomus hydrophilus (Coleoptera: Bathysciinae). *Ecography* **1992**, *15*, 12–18. [CrossRef]

11. Camp, C.D.; Wooten, J.A.; Jensen, J.B.; Bartek, D.F. Role of temperature in determining relative abundance in cave twilight zones by two species of lungless salamander (family Plethodontidae). *Can. J. Zool.* **2014**, *92*, 119–127. [CrossRef]

12. Lunghi, E.; Manenti, R.; Ficetola, G.F. Cave features, seasonality and subterranean distribution of non-obligate cave dwellers. *PeerJ* **2017**, *5*, e3169. [CrossRef]

13. Nitzu, E.; Nae, A.; Băncilă, R.; Popa, I.; Giurginca, A.; Plăiaşu, R. Scree habitats: Ecological function, species conservation and spatial-temporal variation in the arthropod community. *Syst. Biodivers.* **2014**, *12*, 65–75. [CrossRef]

14. Pellegrini, T.G.; Ferreira, R.L. Are inner cave communities more stable than entrance communities in Lapa Nova show cave? *Subterr. Biol.* **2016**, *20*, 15–37. [CrossRef]

15. Lunghi, E.; Manenti, R.; Ficetola, G.F. Seasonal variation in microhabitat of salamanders: Environmental variation or shift of habitat selection? *PeerJ* **2014**, *3*, e1122. [CrossRef] [PubMed]

16. Di Russo, C.; Carchini, G.; Rampini, M.; Lucarelli, M.; Sbordoni, V. Long term stability of a terrestrial cave communty. *Int. J. Speleol.* **1997**, *26*, 75–88. [CrossRef]

17. Poulson, T.L.; Lavoie, K.H.; Helf, K. Long-term effects of weather on the cricket (Hadanoecus subterraneus, Orthoptera, Raphidiophoridae) guano community in Mommouth cave National Park. *Am. Mid. Nat.* **1995**, *134*, 226–236. [CrossRef]

18. Badino, G. Cave temperatures and global climatic change. *Int. J. Speleol.* **2004**, *33*, 103–114. [CrossRef]

19. Scheneider, K.; Christman, M.C.; Faga, W.F. The influence of resource subsidies on cave invertebrates: Results from an ecosystem-level manipulation experiment. *Ecology* **2011**, *92*, 765–776. [CrossRef] [PubMed]

20. Salvidio, S.; Lattes, A.; Tavano, M.; Melodia, F.; Pastorino, M. Ecology of a Speleomantes ambrosii population inhabiting an artificial tunnel. *Amphib. Reptil.* **1994**, *15*, 35–45.

21. Isaia, M.; Giachino, P.M.; Sapino, E.; Casale, A.; Badino, G. Conservation value of artificial subterranean systems: A case study in an abandoned mine in Italy. *J. Nat. Conserv.* **2011**, *19*, 24–33. [CrossRef]

22. Istituto Superiore per la Protezione e la Ricerca Ambientale (2005). Carta Geologica d'Italia alla scala 1:50.000. Available online: http://www.isprambiente.gov.it/Media/carg/213_GENOVA/Foglio.html (accessed on 7 December 2019).

23. Salvidio, S.; Oneto, F.; Ottonello, D.; Pastorino, M.V. Monitoraggio a lungo termine del geotritone Speleomantes strinatii nella Stazione Biospeleologica di San Bartolomeo di Besolagno (GE). In *Atti 22° Congresso Nazionale di Speleologia*; De Nitto, L., Maurano, F., Parise, M., Eds.; Memorie dell'Istituto Italiano di Speleologia: Bologna, Italy, 2015; Volume II, pp. 29, 422–427.

24. Pastorino, M.V.; Pedemonte, S. Nuove stazioni di raccolta del geotritone nell'Oltregiogo Genovese. *Atti XI Congresso Nazionale di Speleologia* **1974**, *2*, 81–82.

25. Ravbar, N.; Kosutnik, J. Variations of karst underground air temperature induced by various factors (Cave of Županova jama, Central Slovenia). *Theor. Appl. Climatol.* **2014**, *116*, 327–341. [CrossRef]

26. Lindström, L.; Reeve, R.; Salvidio, S. Bayesian salamanders: Analysing the demography of an underground population of the European plethodontid Speleomantes strinatii with state-space modelling. *BMC Ecol.* **2010**, *10*, 4. [CrossRef] [PubMed]

27. Salvidio, S.; Oneto, F.; Ottonello, D.; Pastorino, M.V. Lagged influence of North Atlantic Oscillation on population dynamics of a Mediterranean terrestrial salamander. *Inter. J. Biomet.* **2016**, *60*, 475–480. [CrossRef] [PubMed]

28. Salvidio, S.; Pastorino, M.V. Spatial segregation in the European plethodontid salamander Speleomantes strinatii in relation to age and sex. *Amphib. Reptil.* **2002**, *23*, 505–510.

29. White, G.C.; Anderson, D.R.; Burnham, K.P.; Otis, D.L. *Capture-Recapture Removal Methods for Sampling Closed Populations*; Los Alamos National Laboratory 8787 NERP: Los Alamos, NM, USA, 1982; pp. 1–235.

30. Lanza, B. Speleomantes strinatii (Aellen, 1958). In *Fauna d'Italia. 42, Amphibia*; Lanza, B., Andreone, F., Bologna, M.A., Corti, C., Razzetti, E., Eds.; Edizioni Calderini: Bologna, Italy, 2007; pp. 152–156.

31. Salvidio, S. Life history of the European plethodontid salamander Speleomantes ambrosii. *Herpetol. J.* **1993**, *3*, 55–59.

32. Gayanilo, F.C., Jr.; Sparre, P.; Pauly, D. *The FAO-ICLARM Stock Assessment Tools (FiSAT) User's Guide*; Food and Agricolture Organisation of the United Nations: Rome, Italy, 1996; pp. 1–126.

33. Hammer, Ø.; Harper, D.A.T.; Ryan, P.D. PAST: Paleontological statistics software package for education and data analysis. *Palaeontol. Electron.* **2001**, *4*, 9. Available online: http://palaeo-electronica.org/2001_1/past/issue1_01.htm (accessed on 21 October 2019).

34. ter Braak, C.J.F.; Verdonschot, F.M. Canonical correspondence analysis and related multivariate methods in aquatic ecology. *Aquat. Sci.* **1995**, *57*, 255–289. [CrossRef]

35. Legendre, P.; Legendre, L. *Numerical Ecology*, 2nd ed.; Elsevier Science: Amsterdam, The Netherlands, 1998; pp. 612–616.

36. Mammola, S.; Isaia, M. Cave communities and species interactions. In *Cave Ecology*; Ecological Studies 235; Moldovan, O.T., Kováč, L., Halse, S., Eds.; Springer Nature: Cham, Switzerland, 2018; pp. 255–267.

37. Sket, B. Can we agree on an ecological classification of subterranean animals? *J. Nat. Hist.* **2008**, *42*, 1549–1563. [CrossRef]

38. Bellés, X. Survival, opportunism and convenience in the processes of cave colonization by terrestrial faunas. *Oecol. Aquat.* **1991**, *10*, 325–335.

39. Culver, D.C.; Pipan, T. *Shallow Subterranean Habitats: Ecology, Evolution, and Conservation*; Oxford University Press: Oxford, UK, 2014; pp. 188–199.

40. Salvidio, S.; Palumbi, G.; Romano, A.; Costa, A. Safe caves and dangerous forests? Predation risk may contribute to salamander colonization of subterranean habitats. *Sci. Nat.* **2017**, *104*, 20. [CrossRef]

41. Lunghi, E.; Manenti, R.; Ficetola, G.F. Do cave features affect underground habitat exploitation by non-troglobite species? *Acta Oecol.* **2014**, *55*, 29–35. [CrossRef]

42. Ficetola, G.F.; Pennati, R.; Manenti, R. Spatial segregation among age classes in cave salamanders: Habitat selection or social interactions? *Popul. Ecol.* **2013**, *55*, 217–226. [CrossRef]

43. Romero, A. *Cave Biology: Life in Darkness*; Cambridge University Press: Cambridge, UK, 2009; pp. 1–291.

44. Romero, A. Caves as biological spaces. *Polymath* **2012**, *2*, 1–15.

45. Mammola, S. Finding answers in the dark: Caves as models in ecology fifty years after Poulson and White. *Ecography* **2019**, *42*, 1331–1351. [CrossRef]

MDPI

St. Alban-Anlage 66

4052 Basel

Switzerland

Tel. +41 61 683 77 34

Fax +41 61 302 89 18

www.mdpi.com

Diversity Editorial Office

E-mail: diversity@mdpi.com

www.mdpi.com/journal/diversity